The iPad Project Book

Stuff you can do with your iPad

MICHAEL E. COHEN
DENNIS R. COHEN
LISA L. SPANGENBERG

The iPad Project Book
Michael E. Cohen, Dennis R. Cohen, and Lisa L. Spangenberg

Peachpit Press
1249 Eighth Street
Berkeley, CA 94710
510/524-2178
510/524-2221 (fax)

Find us on the Web at: www.peachpit.com
To report errors, please send a note to errata@peachpit.com.

Peachpit Press is a division of Pearson Education.
Copyright © 2011 by Michael E. Cohen, Dennis R. Cohen, and Lisa L. Spangenberg

Executive editor: Clifford Colby
Editor: Kathy Simpson
Production editor: Danielle Foster
Compositor: Danielle Foster
Indexer: Ann Rogers
Cover design and photo compositing: Mimi Heft
Interior design: Peachpit Press

Notice of Rights
All rights reserved. No part of this book may be reproduced or transmitted in any form by any means, electronic, mechanical, photocopying, recording, or otherwise, without the prior written permission of the publisher. For information on getting permission for reprints and excerpts, contact permissions@peachpit.com.

Notice of Liability
The information in this book is distributed on an "As Is" basis without warranty. While every precaution has been taken in the preparation of the book, neither the author nor Peachpit shall have any liability to any person or entity with respect to any loss or damage caused or alleged to be caused directly or indirectly by the instructions contained in this book or by the computer software and hardware products described in it.

Trademarks
Apple, iPad, iTunes, iPhone, iPod, and Mac are trademarks of Apple, Inc., registered in the United States and other countries.

Many of the designations used by manufacturers and sellers to distinguish their products are claimed as trademarks. Where those designations appear in this book, and Peachpit was aware of a trademark claim, the designations appear as requested by the owner of the trademark. All other product names and services identified throughout this book are used in editorial fashion only and for the benefit of such companies with no intention of infringement of the trademark. No such use, or the use of any trade name, is intended to convey endorsement or other affiliation with this book.

ISBN-13 978-0-321-71475-6
ISBN-10 0-321-71475-X

9 8 7 6 5 4 3 2

Printed and bound in the United States of America

Michael and Lisa: For the late Vinton Dearing, who would be astonished to see what computers can do with text today.

Dennis: To my wonderful wife, Kathy, who might love her iPad even more than I love mine (hard to believe, but possible). Also to our kids, grandkids, great-grandkids, and four-legged family members (especially Spenser and Maggie).

About the Authors

Michael E. Cohen has been (in no particular order) a teacher, a programmer, an editor, a short-order cook, a postal clerk, a Web designer, a digital media producer, an instructional-technology consultant, a certified usability analyst, and an assembly-line worker. A three-time contributing editor of *The Macintosh Bible,* he co-wrote the *Apple Training Series: iLife '09* and is the author or co-author of several other books. He lives in Santa Monica, California, with about a half-dozen working Macs and the memory board from his Apple Lisa.

Dennis R. Cohen has been developing software since his days with the Jet Propulsion Lab's Deep Space Network and has been writing and editing books and magazine articles since the late 1970s. He's author, co-author, or contributing author of almost 30 titles and the editor of more than 300 technology titles.

Lisa L. Spangenberg, an expert in medieval English and Celtic languages, writes about technology, food, and books when she isn't administering Web servers and creating Web sites. She has wanted an iPad since 2000.

For more information about the authors—and about all things iPad—see their Web site at www.ipadprojectsbook.com.

Acknowledgments

Michael would like to express his thanks to Cliff Colby, who turned a series of phone conversations into a real live book, and to Kathy Simpson, who edited his discursive ramblings into something resembling coherent prose.

Dennis would like to thank Apple for creating hardware and software that is not only incredibly useful, but also a joy to use. Also, thanks go to the purveyors of the third-party software that so wonderfully enhances Apple's products—in particular, the iPad. Thanks, too, to Michael and Lisa for providing excellent collaboration on a really enjoyable title.

Lisa would like to thank Mac for food and fresh air, Michael for spiffy introductions and practical advice, and Kathy for making me look much better than I deserve.

Contents

Introduction ix

CHAPTER 1 Living in the iPad

Information Syncing Project.. 2
    View your sync settings.................................... 3
    Sync contacts... 6
    Sync calendars... 7
    Sync mail settings....................................... 8
    Sync notes and bookmarks............................. 9
    Apply your sync settings.............................. 10
Wireless Syncing Project 11
    Get a MobileMe account............................... 12
    Go from computer to cloud 13
    Cut the cord 15
    Go from cloud to iPad 16
Simple Security Project.. 18
    Fasten the passcode lock............................. 19
    Protect the young and the restless 21
    Search and recover with Find My iPad............ 23
Mail Management Project 26
    Add contacts 27
    Draft an email 34
    Subscribe to calendars via email.................... 35
    Manage your mailboxes 36
Contact and Calendar Management Project 44

Sort your contacts.................................... 45
Set a calendar... 46
Hear calendar alerts.................................. 48
Use Time Zone Support............................... 49
Get directions... 51

CHAPTER 2 **Working and Playing in the iPad**

File Management Project................................ 54
   Email files to yourself................................ 54
   Transfer a file from a computer....................... 60
      Import files into the iPad........................ 62
   Export files from the iPad............................ 64
   Use Dropbox... 67
   Use Documents To Go – Office Suite.................. 69
iPad Chef Project.. 72
   Find recipes with Epicurious......................... 72
   Find recipes with BigOven Lite....................... 80
   Create a recipe scrapbook............................ 87
Party Project.. 98
   Use Notes to make lists.............................. 98
   Make a shopping list................................ 101
   Create your invitations............................. 104
   Handle the responses............................... 118
Flash Card Project..................................... 119
   Get your apps in gear............................... 119
   Translate some words and phrases................... 121
   Collect some illustrations.......................... 123
   Create your flash-card deck......................... 125
Vacation Planning Project.............................. 132
   Pack your apps..................................... 133
   Set up a trip with TravelTracker..................... 134
   Find flights with KAYAK............................. 141
   Add flight information to your itinerary............. 146

CHAPTER 3 Music, Books, and Movies on the iPad

Music Syncing Project 152
    Sync everything 152
    Sync artists and genres 154
    Make and sync playlists for your iPad 156
    Create a playlist on your iPad. 161
    Manage your music by hand 163
Create and Convert E-Books Project 164
    Make your own PDFs 165
    Convert existing e-books. 166
Movie and TV-Show Syncing Project. 178
    Sync movies 179
    Sync TV-show episodes. 183
Streaming Internet Video Project 187
    Get the software 188
    Use the ABC Player. 190
    View the video stream 194
Streaming Your Video Project 196
    Get Air Video 197
    Introduce your iPad to Air Video Server 199
    Play your content. 203
Watching Television Project. 206
    Get the software, hardware, and app 207
Converting Video Project 213
    Acquire HandBrake and VLC 214
    Convert and transfer your media. 215

Introduction

We saw our first iPad in a theater in Hollywood, California, in the summer of 1968. It appeared in the movie *2001: A Space Odyssey*, and the iPad (called a Newspad in the Arthur C. Clarke novel on which the movie is based) made its debut when astronaut Dave Bowman used it to view the news while having a horrific-looking meal of puréed space food. We didn't want any of that food, but boy, did that Newspad look appetizing.

It took only 42 years (interesting number, 42) for the iPad to make it from Hollywood to the Apple Store.

Even more entertaining than the movie was the consternation and confusion among technology pundits when the iPad was announced in 2010. Very few of them could figure out what the device was *for,* and all too many of them were convinced that it wouldn't be popular.

The iPad turned out to be very popular, and the public immediately figured out what it was for.

What the iPad Is For

What *is* the iPad for? It's for fun. It's for work. It's for convenience. It's for doing whatever a legion of app developers can make a sleek, bright, big-screen, handheld, touch-driven device do—reading books, playing games, looking at photos, looking up at the stars, doing budgets, sending and receiving email, browsing the Web, reserving plane tickets, watching movies or TV, listening to music, writing novels or sonnets, drawing pictures, and so on.

What This Book Is For

A better question is: What is this *book* you're reading for? It's for showing you how to take advantage of your sleek, bright, big-screen, handheld, touch-driven device.

We call it *The iPad Project Book* because we present this information in the form of projects: simple tasks that you can complete in a few minutes each and that reveal much of your iPad's hidden splendor.

Some projects walk you through basic procedures, like getting your music synced between your computer and your iPad. Other projects help you do fun and useful things with your iPad, like planning a vacation and getting flight reservations.

We've divided the book into the following three chapters:

- **Living in the iPad.** This chapter contains projects that help you perform basic tasks on the iPad, such as syncing your contacts and calendars, setting up security, and handling your mail.

- **Working and Playing in the iPad.** This chapter shows you how to do stuff, such as plan a party or a vacation, and make stuff, such as create a deck of flash cards for learning another language.

- **Music, Books, and Movies on the iPad.** This chapter contains projects for navigating the various e-book applications you can put on an iPad; putting music, movies, and videos on the iPad; and creating e-books to read on your iPad.

This book only scratches the surface of what you can do with your iPad. After all, it's a magical device, and there's a lot you can do with magic.

A Note About Conventions

Unlike most computer books, this one is short on technical terminology and conventions. Still, there are a few things you need to know.

To begin, even though the iPad works with both Windows PCs and with Macs, all the authors are Mac users. Therefore, the screen shots we provide from computers are from Macs. PCs and Macs are looking more alike all the time, however, so we don't think these screen shots will be a problem for our Windows-using readers.

Also, we tend to use Mac terminology, referring to *dialogs* instead of *dialog boxes* and *pop-up menus* instead of *drop-down lists*. We're sure that you can figure things out, and we do note when Windows and Mac instructions differ (fortunately, fairly seldom).

Sometimes, we say things like "Tap Settings > General > Network." This is a shortcut way of saying, "Tap the Settings app. When Settings opens, tap General, and then, in the General screen, tap Network." Again, we're sure that you can figure this out.

Finally, you need to know a few basic iPad action terms:

- **Tap.** This means touch your finger to the screen and then quickly lift it.

- **Tap and hold.** This means touch the screen and *don't* lift your finger.

- **Swipe.** This means touch the screen and quickly drag your finger up, down, left, or right. (We tell you the direction in which to drag.)

And now, with that out of the way, on to the projects.

1

Living in the iPad

If you have an iPad (and surely you do, because otherwise, you wouldn't have bought this book), chances are very good that you have another computer—if not two, or three, or more—knocking about your home or office. What's more, you probably have a bunch of information on your computer that you'd find very useful to have on your iPad as well: contacts, calendars, bookmarks, mail settings, and so on.

This street isn't one-way: As you live with your iPad, you'll find yourself putting stuff on it that you'll want to transfer to your computer.

The projects in this chapter show you how to get your stuff from here to there and back again, as well as a few cool things you can do with that stuff when you share it among your devices.

Information Syncing Project

Difficulty level: Easy

Software needed: iTunes

Additional hardware: None

The act of getting your information from your computer to your iPad and back is called *syncing* (short for *synchronizing*), and it's more than just a simple matter of copying your stuff from your computer to your iPad or from your iPad to your computer.

Syncing involves looking at two similar sets of information (such as the contacts in your address book on your computer and the contacts on your iPad), figuring out what's different between those two sets, and sorting things so that the differences between those two sets of information are resolved. Contacts that you created on your iPad go to your computer, for example; contacts that you created on your computer go to your iPad; and contacts that you changed on one device or the other are brought into alignment.

In Apple's world of handheld devices, the key to getting your stuff from here to there and back again is iTunes.

Yes, we know—things like contact lists, appointment books, and browser bookmarks aren't songs, so it does seem a little odd (OK, *more* than a little odd) to use iTunes to move them back and forth between your computer and your iPad. Don't question. That's just the way it is. Embrace it.

So how do you sync your iPad and your computer? Simple: Connect them with the dock connector, and stand back. Unless you've fiddled around with the default settings, iTunes opens and automatically syncs the two devices. It does this each and every time you connect your iPad to your computer.

Any time you don't want iTunes to sync your iPad and your computer automatically, you can hold down the Option and Command keys (Mac) or the Shift and Ctrl keys (Windows) when you connect your iPad. Keep holding those keys down until your iPad appears in the iTunes Source list.

Why iTunes for Syncing?

If you *really* want to know why you use iTunes to sync so much nonmusic stuff, the answer has to do with history and evolution.

In the beginning, there was iTunes, which stored and played your music for you. Next came the iPod, a music-playing device, and it seemed only natural for Apple to use iTunes as the software that moved music from your computer to that device. Also, because the iPod had a screen that could display text, Apple provided—just as an extra-special bonus—the ability to copy contacts and calendars from your computer to your iPod so that you could have them with you as you walked around listening to your music.

But the iPod soon developed more capabilities, such as the ability to show video, so Apple added video playback to iTunes and enabled iTunes to share that video with the iPod.

Then came the iPhone, which was like an iPod from the future: It could not only play music and video and display text, but also handle email, create appointments and contacts, browse the Web, and run applications. So Apple gave iTunes the ability to sync contacts and calendars and apps and bookmarks between the iPhone and the computer.

After that came the iPad, which can handle even more kinds of information, and Apple grafted the ability to sync those kinds of information onto iTunes as well.

That's where we are today, with the distant descendant of the original music-playing application managing all sorts of information on the distant descendant of the original handheld music-playing device: evolution and intelligent design joining hands, wearing white earbuds, and dancing together. We get all misty just thinking about it.

 If you never want your iPad to sync with your computer when you connect it, open iTunes Preferences (choose iTunes > Preferences on a Mac or Edit > Preferences in Windows), click Devices, and select Prevent iPods, iPhones, and iPads from syncing automatically.

View your sync settings

Syncing isn't much of a project, however. You can train a helper monkey (or a bright toddler) to do that much. Instead, in this project you customize the sync settings for your contacts, calendars, email, bookmarks, and more so that only the information that you want to sync gets synced.

note Your iPad can sync with a lot of sources in a lot of ways—directly over the air with MobileMe, for example, or with Google. You can mix and match syncing methods, but the number of combinations can become complex. In this project, we're going to go with the simplest case: syncing by way of direct connection between your iPad and your computer.

First, you connect your iPad to your computer and confirm a general option that controls how your iPad and iTunes interact.

Connecting your iPad and viewing general options:

1. Connect your iPad to your computer with the dock connector.

   If you haven't changed any of the default settings for the iPad or iTunes, iTunes begins syncing with your iPad after a few seconds, and the iPad appears in the iTunes Source list—the sidebar below the Devices heading (**Figure 1.1**). Go ahead and let it sync; that won't hurt anything.

Figure 1.1 The iPad appears in the iTunes Source list when you connect it to your computer.

Source list

Chapter 1: Living in the iPad   5

> **note** In addition to syncing, iTunes backs up the information on your iPad every time you connect it. This backup takes place before any information is synced.

2. Select your iPad in the iTunes Source list.

   In the main iTunes pane, a Summary tab appears (**Figure 1.2**), displaying information about your iPad and a variety of options.

Figure 1.2 The iTunes Summary tab for an iPad is more than just a summary; it also has options you can set.

3. Make sure that the check box titled Open iTunes when this iPad is connected is checked.

   This setting is the best one to use if you want to sync your iPad with iTunes regularly.

The other options aren't relevant for syncing your general information with your iPad. You can ignore them for now (but not forever).

The real fun stuff is on the next tab to the right in the main iTunes pane: Info. That tab is where you set up syncing for your contacts, calendars, bookmarks, mail settings, and notes.

Sync contacts

First, you'll set up contact syncing. iTunes knows about various contact sources on your computer and on the Internet, and it allows you to pick which ones to sync with your iPad, depending on your operating system:

- **On a Mac,** you can sync from Address Book as well as from other contact sources, such as Yahoo! Address Book and Google Contacts.

- **In Windows,** you can sync from only one source of contacts at a time. Your options include Yahoo! Address Book, Google Contacts, Windows Address Book (Microsoft Outlook Express), Windows Vista Contacts, and Microsoft Outlook 2003 or 2007.

On both a Mac and a Windows PC, you can organize your contacts in groups. iPad contact syncing allows you to sync only specific groups of contacts, if you like.

Setting contact sync options:

1. Make sure that the iPad is connected to your computer, that iTunes is open, and that the iPad is selected in the iTunes Source list.

    This will be the case if you just completed the steps in "Connecting your iPad and viewing general options" earlier in this project.

2. In the main iTunes pane, click the Info tab.

    Near the top of the Info tab, a panel displays the contact syncing options. The options in this panel differ, depending on whether you have a Mac or a PC running Windows.

3. Do one of the following:

    - *On a Mac,* check Sync Address Book Contacts (**Figure 1.3**).

    - *In Windows,* check Sync contacts from and then choose the source of the contacts that you want to sync from the menu. Depending on the source you choose, you may have to enter login credentials so that iTunes and your iPad can access the contacts.

Figure 1.3 The contact syncing options on the Mac allow you to sync from several contact sources at the same time.

4. If you want to sync only specific contact groups, click Selected groups and then check the contact groups that you want to appear on your iPad; otherwise, click All contacts.

5. If you want the new contacts you create on your iPad to belong to a specific contact group, check the box titled Add contacts created outside of groups on this iPad to; then choose a group from the pop-up menu.

6. (Optional) On a Mac, check Sync Yahoo! Address Book contacts and then enter your Yahoo login information.

7. (Optional) Also on a Mac, check Sync Google Contacts and then enter your Google login information.

8. In the bottom-right corner of the Info tab, click Apply.

    iTunes applies the changes you made and syncs your iPad.

Sync calendars

With your contact sync settings squared away, next you'll set up the calendars that you want to sync, depending on your operating system:

- **On a Mac,** you can sync your iCal calendars, which can include calendars from any application that syncs with iCal, such as Microsoft Entourage.

- **In Windows,** you can choose to sync calendars with specific versions of Microsoft Outlook.

Setting calendar sync options:

1. In the Info tab, do one of the following:

   - *On a Mac,* check Sync iCal Calendars (**Figure 1.4**).

   - *In Windows,* check Sync calendars with; then, from the drop-down menu, choose the application that manages the calendars you want to sync.

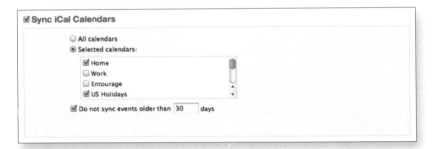

Figure 1.4 The calendar syncing options allow you to choose specific calendars.

2. If you want to sync only specific calendars, click Selected calendars and then check the calendars that you want to appear on your iPad; otherwise, click All calendars.

3. (Optional) Click Do not sync events older than *x* days and then enter a number in the text box.

Unless you think you'll need to refer to past events on your iPad, choosing not to sync events older than 30 days (the default) helps keep your iPad calendar uncluttered.

Sync mail settings

First, let's be clear: iTunes does *not* sync mail between your iPad and your computer. Instead, both your computer and your iPad obtain and display your email directly from your email provider (or providers; Michael currently has email accounts with four different providers, for example).

What iTunes *does* sync are the settings you've enabled for each of your email accounts. This feature is a shortcut, really: It helps you skip the sometimes-confusing task of specifying settings for each email account manually on your iPad by copying those settings directly from your computer.

 **Apple calls this process syncing, but the syncing goes only one way: from your computer to your iPad. If you change email settings on your iPad, they don't sync back to your computer.**

Although a plethora of email programs is available for both Mac and Windows, iTunes can sync settings from only a few of those programs—specifically, Mail on a Mac and Microsoft Outlook (2003 or 2007) or Outlook Express on a PC.

If you have and use any of the iTunes-blessed mail programs, you can quickly get your mail settings onto your iPad.

Syncing mail settings:

1. In the Info tab, check Sync Mail Accounts (**Figure 1.5**).

2. In the list of mail accounts, check each account that contains settings you want to sync to your iPad.

Figure 1.5 Choose the mail-account settings you want to copy to your iPad.

Sync notes and bookmarks

The iPad has a built-in browser (Mobile Safari) and a note-taking app (Notes), and you can sync bookmarks and notes from these apps between your iPad and your computer, based on your operating system:

- **On a Mac,** you can sync your bookmarks between Mobile Safari on the iPad and Safari. (Sorry, Mozilla Firefox fans—no soup for you.)

- **In Windows,** you can sync bookmarks between your iPad and either Safari or Internet Explorer. (*Still* no soup for you, Firefox fans.)

The notes you create on your iPad can sync to the Mail application on the Mac or to Microsoft Outlook in Windows, and vice versa.

Syncing notes and bookmarks:

1. In the Other section of the Info tab, check the bookmark syncing option you prefer:

    - *On a Mac,* your only choice is Safari.

    - *In Windows,* you can choose Internet Explorer or (if it's installed) Safari.

2. Check the notes-syncing option.

From now on, whenever you make or change a note on your iPad, or add a bookmark, the info gets synced to your computer. Similarly, any notes or bookmarks that you add or change on your computer (if you're using one of the iTunes-supported programs) appear on your iPad the next time you sync.

Apply your sync settings

Finally, it's time to apply the settings that you specified and get your iPad resynced the way you want it.

Applying sync-settings changes:

1. In the bottom-right corner of the Info tab, click Apply.

    iTunes applies the changes you made and syncs your iPad.

2. When the sync is complete, disconnect your iPad.

    The iPad disappears from the iTunes Source list, and you're ready to go enjoy your newly synced iPad.

In the future, whenever you connect your iPad to your computer, the sync settings you made in this project are in effect, syncing just the information you want to sync between your iPad and your computer.

Wireless Syncing Project

Difficulty level: Moderate

Software needed: MobileMe account

Additional hardware: None

When you begin living with your iPad, you soon discover that it's really convenient to put a lot of the personal information that you keep on your desktop or laptop computer on your iPad as well. The preceding **Information Syncing Project** shows how to bring all that stuff into alignment every time you connect your iPad to your computer.

But using that method of syncing your information means that you *have* to connect your iPad to your computer every so often to keep each device up to date, and for best results, you need do this regularly and frequently. It's just like brushing your teeth.

We live, however, in an age of miracles and wonder—and of wireless technology. There's no reason why your iPad and your computer can't share their information over the air so you can avoid the daily chore of getting them all synced up. (You should still brush your teeth, of course.)

No matter whether you have a Wi-Fi or 3G iPad, you can make use of cloud services to sync your information between your computer and your iPad. (To find out what we mean by *cloud services,* see the nearby sidebar.) This project describes how to use a cloud service offered by Apple: MobileMe. In this project, you'll sync your contacts, calendars, and bookmarks with MobileMe from both your computer and your iPad.

What Is This Cloud of Which You Speak?

Cloud is one of those terms that has moved out of the cloistered halls of geekdom into the light of day. It really means nothing more than *somewhere on the Internet, outside your local network.* The term comes from the cloud-shaped symbol used in the network diagrams that networking professionals draw from time to time when they get tired of reading router logs. Like a real-world cloud, a virtual cloud is a shapeless object—one into which you can't easily see.

In this project, *cloud* refers to the Apple servers on which your iPad information is stored so that you can get to it from anywhere on the Internet.

If you already have a MobileMe account and use it to sync your contacts, calendars, and bookmarks between your computer and MobileMe, you can skip the next two sections and go right to "Cut the cord." Otherwise, read on.

Get a MobileMe account

To use a cloud-based service, you need to have an account with that service, such as Apple's MobileMe.

A MobileMe account is free for the first 60 days so that you can try out the service; it costs $99 a year after that. If you have a Mac, Apple makes it very easy to get a MobileMe account; if you don't already have one, you've probably been asked by your Mac to set one up on more than one occasion.

Sync Outside the Box

MobileMe isn't your only wireless syncing option for your iPad, of course. You can use one of these services instead:

- **Google.** Google supports wireless syncing with its services. You can find out more at the Google Sync page (www.google.com/support/mobile/bin/topic.py?hl=en&topic=14252).

- **Microsoft Exchange.** Readers who use a Microsoft Exchange server for school or business aren't left out of the party either, because the iPad can sync with an Exchange account. Although the setup process usually is simple, you should consult the Exchange server's administrator to see whether it's configured to support iPad users.

The initial release of the iPad software supports only one Exchange account on your device, although the next version of the iOS for iPad will support multiple Exchange accounts. That software may be available by the time you read this book.

Signing up for MobileMe:

- **On a Mac,** open System Preferences, click the MobileMe icon, and follow the instructions there.

- **In Windows,** go to www.apple.com/mobileme, and click the prominently displayed Sign up for MobileMe Free Trial button. When you create a MobileMe account in Windows, Apple provides a MobileMe control panel for your Windows system so that you can control the service's various features.

 MobileMe's features—both on the Mac and in Windows—include email; contacts; calendars; gigabytes of file storage on Apple's servers; and quite a bit more, including a valuable Find My iPhone/iPad service that you can use to locate your device, should you lose it.

Go from computer to cloud

Now that you have a MobileMe account, you need to sync your information between it and your computer so that your iPad can get your information from MobileMe:

- **On a Mac,** you can sync your contacts from Address Book, your calendars from iCal, and your bookmarks from Safari.

- **In Windows,** you can sync your contacts from Microsoft Outlook 2003 or 2007, from Vista Contacts, or from Windows Address Book. You sync your calendars from Outlook; if you use another calendar program, you need to move your calendar events to Outlook if you want to sync them with MobileMe. You can sync your bookmarks from either Internet Explorer or Safari. (Yes, Apple offers a Windows version of Safari.)

 After you sync your information between your computer and MobileMe, you can access it from any computer with a modern Web browser. Just go to www.me.com and log in.

Syncing from a Mac with MobileMe:

1. Open System Preferences, and click MobileMe.

2. In the MobileMe window, click the Sync tab.

3. Select the Synchronize with MobileMe check box, and choose Automatically from the adjacent pop-up menu.

    When you choose Automatically, your Mac and MobileMe will sync information as soon as you change it.

4. In the list of items that you can sync, select Bookmarks, Calendars, and Contacts (**Figure 1.6**).

Figure 1.6 The Sync tab of the MobileMe System Preferences window in Mac OS X.

5. Click the Sync Now button.
6. Close System Preferences.

   Your Mac sends your information to the cloud, ready to be synced with your iPad (or with any other computers and devices that you sync with the same MobileMe account).

Syncing from a Windows PC with MobileMe:

1. Choose Start > Control Panel to open Control Panel.
2. In the Network and Internet section, choose the MobileMe control panel.
3. Sign in with the MobileMe member name and password that you created when you set up your MobileMe account.
4. Click the Sync tab.

5. Select Sync with MobileMe, and choose Automatically from the drop-down menu.

6. Select Contacts, and choose the application that manages your contacts from the drop-down menu.

7. Select Calendars, and choose the calendar application from the drop-down menu.

8. Select Bookmarks, and choose the Web browser that you use from the drop-down menu.

9. Click the Sync Now button.

10. Close the MobileMe control panel.

Cut the cord

Reading this section is necessary only if you've completed the **Information Syncing Project** earlier in this chapter. In this section, you undo what you did in the earlier project: You turn off wired syncing of contacts, calendars, and bookmarks between your iPad and your computer.

Most likely, nothing will go wrong if you don't turn off wired syncing of your information. When you sync the same information both over a dock-connector cable with iTunes and wirelessly with MobileMe, however, you might end up with duplicate information.

Turning off wired syncing of your information:

1. Launch iTunes, if it's not already running.

2. Connect your iPad, and let the sync take place.

    After the sync, your iPad and your computer have the same information. What's more, your computer and MobileMe should also be in sync if you set up MobileMe to sync automatically (as we show you how to do in "Go from computer to cloud" earlier in this project).

3. In the iTunes Source list, select your iPad.

4. In the main iTunes pane, click the Info tab.

5. Clear the check box titled Sync Address Book Contacts (Mac) or Sync Contacts (Windows).

6. Clear the check box titled Sync iCal Calendars (Mac) or Sync Calendars (Windows).

7. In the Other section, clear the check box titled Sync Safari Bookmarks (Mac) or the corresponding check box in Windows.

    In Windows, the name of this option depends on the browser with which you sync bookmarks. Internet Explorer, for example, calls bookmarks *favorites*, so if you use it to sync bookmarks, this option would be called Sync Favorites.

8. Disconnect your iPad from the dock-connector cable.

Go from cloud to iPad

Now you're ready to set up your MobileMe account on your iPad and activate contact, calendar, and bookmark syncing. If you previously synced these items with iTunes, you'll also specify what to do with the information that remains on your iPad.

Creating a MobileMe account on the iPad:

1. On the iPad, tap the Settings app.

2. In the pane on the left side of the Settings screen, tap Mail, Contacts, Calendars.

3. In the Accounts section of the resulting screen, tap Add Account.

4. In the Add Account screen, tap MobileMe.

    An account-entry form appears (**Figure 1.7**).

5. Tap and type to enter the requested information: your name (the one you normally go by), the email address assigned to your MobileMe account, the password for your MobileMe account, and an optional description of that account.

Chapter 1: Living in the iPad   **17**

Figure 1.7
The MobileMe
account form.

6. Tap the Next button in the top-right corner.

   A second account form appears, in which you can turn MobileMe features on or off for your account.

7. Tap the Contacts switch in the form to turn on Contacts syncing.

   If you have any contacts on your iPad (and you probably do, if you've completed the **Information Syncing Project** earlier in this chapter), a Merge Contacts dialog appears, asking you what to do with any existing contacts on your iPad (**Figure 1.8**).

Figure 1.8 The Merge Contacts dialog allows you to merge your iPad information with MobileMe or to discard it and start fresh.

8. If you see the Merge Contacts dialog, tap Merge with MobileMe.

   If you worked through the preceding sections of this project, the contacts on your iPad already are the same as the ones in MobileMe, so merging should have no effect; identical contacts are ignored.

9. Tap the Calendars switch, and if the Merge Calendars dialog appears, tap Merge with MobileMe.

    Again, if you followed the instructions earlier in this project, the calendars on your iPad should already match those in MobileMe, so merging them has no effect.

10. Tap the Bookmarks switch, and then tap Merge with MobileMe in the Merge Bookmarks dialog that appears.

11. Tap the Save button in the top-right corner of the MobileMe account form.

    The form closes, and you return to the Mail, Contacts, Calendars page.

12. Tap Fetch New Data.

13. In the screen that appears, tap the Push switch to turn it on.

14. Press the Home button on your iPad.

    Now your iPad and MobileMe are set to communicate. Whenever you make a change in your calendars, contacts, or bookmarks on your computer, the change is sent to your iPad over the Internet. Similarly, any changes you make in your contacts, calendars, or bookmarks on your iPad are sent over the Internet to your computer. It's magical!

Simple Security Project

Difficulty level: Easy

Software needed: MobileMe account (optional)

Additional hardware: None

In the 1976 film *Marathon Man,* the malevolent Dr. Christian Szell asks the protagonist, "Is it safe?" If he'd been asking about your iPad, the answer would have to be "Probably not."

As configured right out of the box, iPad is open to anyone: Pick it up, press the Sleep/Wake button, and drag the onscreen slider to the right, and all your iPad's secrets are revealed. Naturally, this revelation isn't much of a problem if you live alone and your iPad stays at home with you. (Trust us—your cat doesn't care what's on your iPad.) But the iPad is a very portable object, and an attractive one, too; it cries out to be picked up, tried out, played with. If you leave it lying around, someone— even someone you know, trust, and love—can easily find it irresistible.

There isn't any way to make the iPad completely impervious to prying eyes. A knowledgeable person with enough time and the right software can crack almost any security scheme. Nonetheless, you can protect your iPad's contents from the casual snooper.

Also, if you share your iPad with your children from time to time (it's a great way to keep kids entertained on long car trips, for example), you can protect them from content that might be too mature for young eyes and ears.

Finally, you can even have the iPad tell you where it is if you misplace it (easy to do; these things love to hide under magazines and pillows) or if someone "accidentally" wanders off with it.

Fasten the passcode lock

The iPad has a built-in passcode lock that you can use to halt casual snoopers dead in their tracks when they wake it up. As soon as a snooper slides the Slide to Unlock slider, he's greeted by a dialog demanding a four-digit passcode. Without that passcode, all the snooper can see is your Lock Screen wallpaper.

Setting a passcode:

1. On the home screen, tap the Settings app's icon.
2. On the left side of the Settings screen, tap General.
3. On the right side of the resulting General screen, tap Passcode Lock.

    The Set Passcode dialog appears (**Figure 1.9** on the next page).

Figure 1.9 Tap out your passcode in this dialog.

4. Tap out a four-digit passcode of your choosing and then repeat it to confirm the passcode.

 Choose a passcode that you can remember easily but that's hard for other people to guess. People who know you might be able to guess the year you were born, for example, but would be less apt to guess the third through sixth digits of your phone number when you were in high school (and no, that's not the passcode we use, but thanks for playing!).

As soon as you enter the passcode, the passcode lock takes effect. By default, the next time you put your iPad to sleep and then wake it up, you must enter the passcode in the Passcode Lock screen—even if you put the device to sleep for only a few seconds. You can change how long the iPad has to be asleep before the passcode lock takes effect, however.

Changing the passcode-lock interval:

1. In the Passcode Lock screen (see the preceding section), tap Require Passcode.

2. In the Require Passcode screen, tap the length of time you want to elapse before the lock kicks in.

3. (Optional) At the top of the Require Passcode screen, tap Passcode Lock; then, in the Passcode Lock screen, tap the Erase Data switch to turn it on.

   This feature guards against those who might abscond with your iPad and then try number after number to unlock it. After ten attempts, your iPad is erased, so the thief has only your valuable iPad, not the possibly even more valuable data that it contains.

> **tip** **If you manage to erase your data accidentally, you can recover it from the automatic backup that iTunes creates each time you sync your iPad with iTunes**—another good reason to sync with iTunes periodically, even if you normally sync your information wirelessly. See the Apple article at http://support.apple.com/kb/HT1414 for more about restoring your iPad.

Protect the young and the restless

As we mention earlier in this project, the iPad is a wonderful child-distraction device. Even pudgy toddler fingers can manage to tap their way around it. But your iPad may contain some items—such as movies, songs, or TV shows—that you really don't want your toddler to see. Neither do you want your 9-year-old daughter wandering into the iTunes Store and buying a few hundred dollars' worth of tunes to die for.

You can set restrictions to guard against the accidental viewing of certain media or the semiaccidental use of certain apps.

Setting up restrictions:

1. Tap Settings > General > Restrictions to open the Restrictions screen.
2. At the top of the screen, tap Enable Restrictions.

   A familiar passcode dialog appears.
3. Tap out a four-digit passcode of your choosing and then repeat it to confirm the passcode.

 This passcode can be the same as your Passcode Lock passcode (refer to "Setting a passcode" earlier in this project), but for better security, you should choose a different one.

After you establish the passcode, the rest of the choices in this screen become available. With them, you can disallow access to certain apps and features, as well as restrict access to certain types of media based on parental ratings (**Figure 1.10**).

Figure 1.10 Restrict access to various apps and types of content here.

4. Set your restrictions:

   - **Purchases, privacy, and Web surfing.** In the Allow section, for example, you can make the iTunes and App stores inaccessible so that guest users can't buy things; disable Location Services for all apps to protect your privacy; and hide the built-in YouTube and Safari apps to keep young, inquisitive eyes from gazing on some of the Internet's seamier sights.

 Disabling Safari doesn't guarantee that iPad users can't get onto the Web; many apps contain built-in browsers that continue to work even when Safari is disabled.

   - **In-app purchases.** You can further protect your pocketbook by turning off In-App Purchases in the Allowed Content section.

- **Movies and TV shows.** In the same Allowed Content section, you can tap Ratings For to specify the national ratings system that the iPad uses when you restrict access to movies and TV shows. You can restrict movie viewing to G and PG movies when the ratings are set for the United States, for example. Switch the Ratings For setting to France instead, and you can restrict movies by the age-level ratings used in that country.

- **Explicit songs and apps.** You can ban music with explicit lyrics as listed in the iTunes Store, and you can hide apps that have age restrictions as listed in the App Store.

This system isn't perfect. Determined children can find their way around seemingly impervious obstacles, so you still need to exercise parental oversight. But the available restriction options do go a long way toward making your iPad more child-friendly.

Search and recover with Find My iPad

As Paul Simon almost sang, losing an iPad is like a window in your heart; everybody sees you're blown apart. By using MobileMe's Find My iPad feature, however, you may be able to find your lost device and close that window. Even if you can't retrieve it, you can still use the feature to lock your iPad with a new passcode or even to wipe out its contents so that your information can't be hijacked and misused.

As should be obvious from the preceding paragraph, you need a MobileMe account to use the Find My iPad feature. If you don't have one, see the **Wireless Syncing Project** earlier in this chapter to learn how to get one and set it up on your iPad. After that, activating the feature takes just a few taps.

Enabling Find My iPad:

1. Tap Settings > Mail, Contacts, Calendars.

2. In the list of accounts on the right side of the Mail, Contacts, Calendars screen, tap your MobileMe account.

   An account-settings dialog appears (**Figure 1.11** on the next page).

Figure 1.11 Find My iPad is linked to your MobileMe account.

3. Tap Find My iPad to set the switch to On.

4. Tap the Done button.

   Now that Find My iPad is activated, you can ask Apple to locate your device for you, which it can do as long as (a) your iPad is awake, (b) your iPad is sleeping within range of a Wi-Fi network connected to the Internet, or (c) your 3G iPad is within range of the mobile-phone network.

 You can also ask Apple to find your iPad with its Find My iPhone service, which works with iPhones, iPod touches, and iPads. You can use this service in a Web browser, or you can download and use the free Find My iPhone app from the App Store.

Finding your iPad with a Web browser:

1. Go to www.me.com, and log in with your MobileMe ID and password.

   The MobileMe service takes you to your MobileMe mail page.

2. In the top-left corner of the MobileMe mail page, click the cloud-shaped Switch Apps icon.

   A set of MobileMe app icons appears, spanning the page.

3. Click the Find My iPhone icon.

4. In the verification page that appears, re-enter your MobileMe password.

   After a few moments, a map appears, pinpointing your iPad's current whereabouts (**Figure 1.12**).

Figure 1.12 Your iPad is hanging out behind a church in Santa Monica, California. Hallelujah!

> **note** The Find My iPhone app works very much like the Web service. You enter your MobileMe information and then use a map interface almost identical to the one in the Web version.

5. Decide what to do about it.

   When you click your iPad's location on the map, you can choose to have your iPad display a message and play a sound for 2 minutes (useful if you've misplaced your iPad around the house; just listen for the sound to track the wandering iPad down).

If, however, your iPad is in some strange, unanticipated place, in the grasp of some foul iPad-snatcher, you can set a new passcode for your iPad (to prevent the evildoer from accessing your information easily) or even erase the information on your iPad (if you suspect that the evildoer is particularly nefarious and dastardly). See "Fasten the passcode lock" earlier in this project for details.

If you do choose to wipe your iPad's information and are fortunate enough to recover the device later, you can use the Restore feature to put the most recent backup of your information back on your iPad. See http://support.apple.com/kb/HT1414 for more information about restoring your iPad. This tip also works if you have to obtain a replacement iPad instead.

Mail Management Project

Difficulty level: Easy

Software needed: Working email account (free or paid), MobileMe account (optional; $99 per year)

Additional hardware: None

The iPad, aside from its many other virtues, is a useful device for reading, responding to, and managing your email at those times when you aren't shackled to your computer (and how sweet those times can be!).

Managing your email on your iPad is much easier, however, if you use the IMAP email protocol. Most Web-mail applications use IMAP (see the sidebar "IMAP and POP Mail Accounts" later in this project for details), which means that your email is synchronized across devices automatically. If you sync your email accounts and contacts via iTunes or MobileMe, you may not even have to add any accounts to your iPad by hand, but you may want to stop right now and check out the **Information Syncing Project** earlier in this chapter to make sure that your account data is the same on your iPad as it is on your other devices.

In the following pages, we use the word *folder* **to refer to what is sometimes called a** *mailbox* **or** *directory***, simply because the iPad icon for mailboxes and mail directories is a folder, one of which is named Inbox. The terms are really equivalents in practical terms.**

Add contacts

These days, it seems that just about everyone has not only home and work addresses and phone numbers, but also mobile-phone numbers, email addresses, Web pages, blogs, Facebook and LinkedIn pages, Twitter accounts, and two or three instant-messaging or chat accounts. It's a lot of contact information to keep track of, and if you're having to keep track of it on several devices that you use in different places, things get more complicated still.

Fortunately, your iPad's Contacts application can help enormously in tracking, sharing, and using the contact information for the people and companies in your life.

The easiest way to add contacts to the Contacts app on your iPad initially is to sync with your contacts on your computer, via iTunes or MobileMe. But Mail has a couple of smart ways to make adding contacts much easier than filling out the Contact screen by hand (though that's always an option too).

The second-easiest way is to add contacts from emails you receive. You can add information to any extant contact or create a new contact based on information in an email, as we show you in the following tasks.

Adding contacts from email address fields:

1. Find an email in your Inbox (or any other mail folder on your iPad) containing an address that you want to add to your Contacts app.

   You're going to create a new contact entry for that address.

2. Tap the person's name or email address in the From, To, or CC field of the email.

You see a floating window similar to the one in **Figure 1.13**. If you tapped the From field, this window is titled Sender; if you tapped the To field, it's Recipient; and if you tapped the CC field, it's CC. Your contact form may have more or less information than the one shown in the figure.

The email address used in the email is already filled in for you. Below the address are two buttons: Create New Contact and Add to Existing Contact.

Figure 1.13 Using the From line in an email to create a new contact.

3. Tap Create New Contact.

   The floating window (Sender, Recipient, or CC) is replaced by the New Contact form (**Figure 1.14**).

 The Contacts app works in part by matching email addresses, so if you have a contact who has more than one email address, you may accidentally create a second contact for that person. When in doubt, tap the Add to Existing Contact button instead of Create New Contact. You'll see a list of all your contacts in a floating window. If the contact doesn't already exist, just tap Cancel, and you'll return to the floating window shown in Figure 1.13, where you can tap Create New Contact.

Figure 1.14
New Contact form.

4. Fill out the First and Last name fields (if your correspondent doesn't use them as part of his or her email address).

5. Tap the blue Done button in the top-right corner to save the new contact information and close the New Contact form.

   You'll see a contact form similar to the one shown in **Figure 1.15** on the next page. You can edit the information in this form, as we show you in the following task.

Figure 1.15
Contact form.

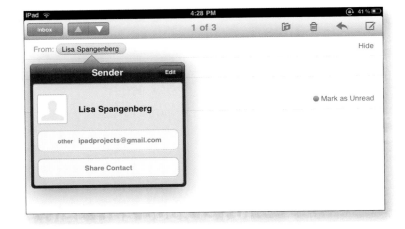

Editing a contact created from an email address:

1. Tap the Edit button in the top-right corner of the contact form (refer to Figure 1.15) to edit or add to the existing information.

   The form expands as shown in **Figure 1.16**.

Figure 1.16 Expanding a contact form.

2. Edit or enter new information in any of the fields.

3. If you want to add an image to the contact, tap the Add Photo rectangle in the top-left corner, and insert an image from the iPad's Photos app.

tip For more projects that use the Photos app, see Chapter 4.

4. Tap the blue Done button in the top-right corner to save your changes and close the form.

Adding contact data from an email body:

1. Find an email with an email address in its body that you want to add to a contact.

note You can also add a street address, phone number, or Web-site URL (see the nearby sidebar "Adding Other Kinds of Data to Contacts from Mail"), but in this task, we're going to show you how to add an email address. The basic procedure for adding all the other types of contact data is the same, though the contact form will look slightly different for each type.

2. Tap and hold the address until you see a floating window similar to the one shown in **Figure 1.17**.

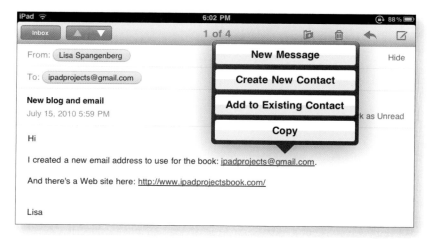

Figure 1.17 Contact options for an address in an email body.

The "tap and hold" part is important. If you merely tap an email address, Mail assumes that you want to send an email immediately, so it creates a blank email form with the tapped address placed in the To field.

3. Tap one of the following buttons to accomplish the associated task:

- **New Message,** which creates a new email for you, already addressed to the email address you tapped
- **Create New Contact,** which opens a New Contact form with the email address already filled in (refer to Figure 1.14 earlier in this project)
- **Add to Existing Contact,** which lets you select and modify an existing contact in the Contacts app
- **Copy,** which places the email address on the iPad's clipboard so that you can paste it into another application or a draft email

Adding Other Kinds of Data to Contacts from Mail

Quite often, someone will send you information in the body of an email that you want to incorporate into a contact, such as a name and address, a phone number, or a Web site's URL. In most cases, Mail recognizes that the http:// and www. prefixes identify a Web site, that phone numbers and street addresses have particular combinations of characters and formatting, and that email addresses use the @ sign. Consequently, when you tap a phone number, a URL, or an email address, Mail tries to predict what you're likely to want to do with it—such as add it to a contact or create a new contact.

Here are a few things you can do with contact data in email:

- If you tap and hold a URL, you see options to open that URL in Safari or copy it to almost anything on your iPad that accepts text, such as a contact or a new email message. If you simply tap a URL, Safari launches and attempts to go to the Web page associated with the URL you tapped.
- If you tap and hold a street address, the options are Open in Maps, which displays a Google street map; Create a New Contact; Add to Contact, which directs Mail to put the address information in the correct fields of the contact form (a task that Mail sometimes has difficulties with); and Copy, which allows you to use the copied address elsewhere. If you simply tap the address, Safari launches and attempts to locate the address in the Maps app.
- If you tap a phone number (domestic or international), you see the options Create New Contact and Add to Existing Contact.

At the bottom of every contact form, just to the right of the Edit button, is a Share button. Tap this button to attach a .vcf (vCard) file to an email. The recipient can automatically add the information in the vCard to his Contacts app or another contact database that supports the .vcf format. We show you how to use this feature in the next task.

Sharing a contact entry:

1. Open an entry in the Contacts app that you want to share via email (**Figure 1.18**).

Figure 1.18 Contact form with Share button visible.

2. Tap the Share button in the bottom-right corner.

    (You may have to scroll down the form to see the Share button, depending on how much information the contact contains.)

Lisa Spangenberg.vcf

You see a blank email form, ready to be addressed, that includes a vCard attachment. The icon for the vCard looks like a small card with a blue human figure.

3. Complete the email form by adding a recipient and a message; then send it.

   Your recipient probably can simply tap (on an iPad) or double-click the icon to add that information to her own contacts list.

Information in the contact's notes field on the iPad isn't included in the shared vCard.

Draft an email

Sometimes when you're writing an email, you need to set it aside to return to later. Mail will save a copy of an unfinished email for you as a draft. You can find it in the special Drafts folder in the account that would have sent the mail, had you finished writing and tapped Send.

Saving a draft email:

1. Tap the gray Cancel button in the top-left corner of the email form.

   You'll see a floating window with Save and Don't Save buttons (**Figure 1.19**).

Figure 1.19 Email draft options.

2. Tap Save.

   The draft is saved to the Drafts folder for that email account.

Mail Shortcuts for Addresses and Photos

We're assuming that you've already set up Mail on your iPad to send and receive email. You've got the basics down; you know how to reply to an email you've received, initiate a new email by tapping the Compose icon, and send your finished email by tapping Send. But the iPad can make emailing easier for you in a couple of ways:

- When you start typing an address in the To field of an email, for example, the iPad, working with the Contacts app, tries to guess the person you're going to send the email to. The results can include people you've corresponded with, even though they aren't listed in your Contacts app.

- If you know exactly whom you want to email, it may be more efficient to tap the blue plus-sign icon ⊕ at the right end of the To field, scroll through your Contacts app, and select that person's email address.

- If you want to email an image to someone, you can use the Photos app, like so:

    1. Tap the Photos app's icon to launch the app on your iPad.

    2. Find the image you want to send, and tap the curved-arrow Export button in the top-right corner.

       You see a form with several buttons, including Email Photo.

    3. Tap Email Photo.

       The iPad switches to Mail, where the image is already attached to a blank email form.

    4. Add an address, subject, and message body.

    5. Send the finished email.

       You're back at the image you selected in step 2.

Subscribe to calendars via email

The iCalendar (.ics) format is supported by lots of calendars besides the Calendar app on the iPad and Apple's iCal. Google Calendar, Basecamp, and Yahoo also support and can use the .ics format.

Sites like iCalShare (http://icalshare.com) and iCal World (www.icalworld.com) make lots of calendars publicly available for free. If you receive an email with a link to a calendar in the body of the email, you can easily subscribe to that calendar in your iPad's Calendar app, as we show you in this section.

Subscribing to a calendar:

1. Open an email that has a link to an .ics-format calendar in the body of the email.

2. Tap the link.

   A dialog like the one shown in **Figure 1.20** opens.

Figure 1.20 Calendar-subscription dialog.

3. Tap Subscribe.

   You see a dialog similar to the one shown in **Figure 1.21**. Tapping the View Events button in this dialog shows you the events in the newly subscribed calendar; the Done button returns you to the email.

Figure 1.21 Confirm the calendar subscription in this dialog.

4. Tap Done.

   You return to the email you were reading in step 1.

You can choose which calendars to display by tapping Calendars in the top-right corner of your iPad's Calendar app and selecting the desired calendars from the list that appears.

Manage your mailboxes

If you have multiple devices that can check email and multiple email accounts, the best mail-management tip we can offer you is to let your email server manage mail for you. Leaving your mail on the server and letting the server be the primary email source means that you'll see the same mailboxes, and find the same messages in the same read or unread states, on all your devices.

Navigating your mailboxes may mean navigating among multiple email accounts. In its initial release, the iPad Mail app has a separate Inbox folder for each account; iOS 4 for the iPad, which may be available by the time you read this book, will provide a single shared Inbox folder for all email accounts.

IMAP and POP Mail Accounts

When you send mail, your email client and ISP use a protocol called SMTP (Simple Mail Transfer Protocol). But when you set up your iPad to receive mail, you usually have the option of telling Mail to use IMAP (Internet Message Access Protocol) or POP (Post Office Protocol) for each email account you set up. In some cases, though, you don't have a choice; your IT department at work or your Internet service provider (ISP) will tell you which protocol to use for incoming mail.

Here are some things to keep in mind about how POP and IMAP work:

- **IMAP** is designed to leave mail on the server and to make sure that every device that logs on to the server has exactly the same mail in the same folders. IMAP accounts let your organize your email in very elaborate ways. You can have lots of folders, and even folders within folders, if you want.

- **POP** is designed to delete email from the server as soon as it has been downloaded to your computer. This arrangement is awkward if you read mail on multiple devices, because you won't see the same incoming emails in all of them. The solution is to make your main computer the one that controls when email is deleted from the server. (A setting in Mail and other email clients tells the server how long to leave incoming mail on the server.) On your other devices that use that POP account, set the time for deleting email to be never—or at least longer than the time you set for your main computer.

You can find a useful explanation of the differences between (and benefits of) POP and IMAP here:

http://docs.info.apple.com/article.html?path=Mail/3.0/en/11920.html

Another challenge associated with email is spam. We live in an era of spam. People we don't know send us email we don't want to read. When you scan your Inbox, one way to isolate the real mail from spam (some legitimate bulk mail sent to mailing lists isn't spam) is to see whether a particular piece of mail was sent to you alone or to you and a bunch of other people. In the following task, we show you how.

Turning on Show To/Cc Label to check for spam:

1. Launch the Settings app on your iPad to open the Settings screen.
2. Tap Mail, Contacts, Calendars in the Settings column.

   You see the pane shown in **Figure 1.22**.

Figure 1.22 The Mail, Contacts, Calendars pane.

3. Tap the switch to the right of Show To/Cc Label to set it to On (**Figure 1.23**).

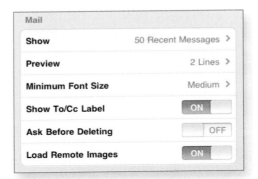

Figure 1.23 Set the Show To/Cc Label switch to On.

From now on, mail in your Inbox and other mail folders addressed *only* to you has a small gray To icon next to it.

Deleting a single email:

1. Open an email that you want to delete.

2. Tap the trash-can icon in the top-right corner of the message.

   The email shrinks and disappears into the trash can in what Apple calls the *genie effect*.

> **tip** If you happen to delete an email by mistake and realize your error immediately (or very soon thereafter), you can recover the email by opening the Trash folder associated with the account in question (by tapping the trash-can icon) and moving the email from the Trash to your Inbox.

> **tip** Depending on your email account settings and the settings on your service provider's mail server, mail that you delete on your iPad may remain in a deleted email or trash folder for a while. Deleting the email in that folder should delete it permanently.

Deleting multiple emails:

1. Tap a mail folder on your iPad.

   You see a floating window similar to the one shown in **Figure 1.24**.

Figure 1.24 A mail-folder list in Mail.

2. Tap the Edit button in the top-right corner of the floating window.

   You see a floating window listing the email in that folder (**Figure 1.25**).

3. Tap the circle to the left of a message that you want to delete.

   The circle turns red with a white check mark in its center (**Figure 1.26**).

Chapter 1: Living in the iPad   **41**

Figure 1.25 Mail-folder list, ready to be edited.

Figure 1.26 Selecting mail for deletion.

4. Repeat step 3 to select as many emails as you want.
5. When you've selected all the emails that you want to delete, tap the red Delete button at the bottom of the floating window.

    The selected emails disappear (via the genie effect) into the top-left corner of the floating window.

Sometimes, you want to keep email in folders other than the Inbox folder. If you've set up other folders for IMAP accounts, including Web-hosted accounts such as Gmail, you can see and use those folders in Mail on your iPad. We show you how in the following tasks.

Moving a single email to a different folder:

1. Open an email that you want to move out of your Inbox.

2. Tap the folder icon at the top of the window.

   You see a floating window listing your mail folders, similar to the one shown in **Figure 1.27** (though likely with very different folders).

Figure 1.27 List of mail folders for a single email account.

3. Tap the folder to which you want to move the open email.

   The email shrinks into that folder via the genie effect.

You can also move several emails from one mail folder to another at the same time, as we show you in the next task.

Moving multiple emails to a single folder:

1. Open a mail folder.

    You see a floating window listing the mail in that folder (similar to the one shown in Figure 1.24 earlier in this project).

2. Tap the Edit button in the top-right corner of the floating window.

    You see a new floating window (similar to the one shown in Figure 1.25 earlier in this project).

3. Select the messages you want to move by tapping them.

    The gray circle to the immediate left of each selected email becomes a red circle with a white check mark in it (**Figure 1.28**).

Figure 1.28 Selecting multiple emails in a list to be moved to a new folder.

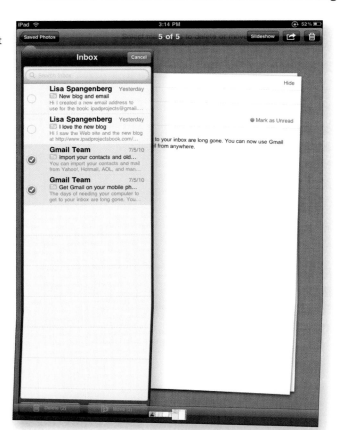

4. Tap the Move button in the bottom-right corner of the floating window.

   You see a list of mail folders similar to the one shown in Figure 1.27 earlier in this project.

5. Tap the destination mailbox or folder in the floating window.

   You see the emails shrink into that folder via the genie effect, and the iPad returns you to your open email.

The button in the top-left corner of an open email displays the name of the email account containing the mail you're viewing. Tapping that button takes you up a level, to the folder that contains the email. At the top level of any account, the button in the top-left corner is labeled Accounts, and tapping it takes you to a list of all the email accounts on your iPad. If you have nested email folders, you may have to tap the button several times to navigate to the top level of that email account's mail folders.

Contact and Calendar Management Project

Difficulty level: Easy

Software needed: None

Additional hardware: None

Other projects show you how to move your information from your computer *to* your iPad; this simple project helps you arrange and view your information in different ways after it's *on* your iPad.

Suppose that you like to look up your contacts by first name, and we like to look up ours by last name. Although this scenario may seem like a case of "You say *po-tay-toe,* and I say *po-tah-toe,*" wars have been fought over less. (No, really! Ask Henry V about tennis balls sometime.) Therefore, in an attempt to foster universal harmony and peace among all people of good will, follow along as we show you the simple settings you can control to tailor the way you interact with your information on your iPad.

Sort your contacts

Our trivial example is actually not so trivial. A business user may want her contacts to be arranged by surname, just as in a phone book or a corporate directory, but someone whose contacts list consists mostly of friends and relatives may well want to find his contacts by those names he knows best: first names.

You can sort your contacts either way. What's more, no matter how you sort them, the iPad offers a way to present those contacts with the first name shown either before or after the last name.

Changing the sort order and presentation of contact names:

1. Tap Settings > Mail, Contacts, Calendars.

2. In the Mail, Contacts, Calendars screen, swipe down until you see the Contacts section (**Figure 1.29**).

Figure 1.29 The sort settings for contacts are near the bottom of the screen.

3. Tap Sort Order to open the Sort Order screen.

4. Tap the sort order that you prefer.

5. At the top of the Sort Order screen, tap Mail, Contacts, Calendars and then tap Display Order.

6. In the resulting Display Order screen, tap the display order that you prefer.

   In case you don't know what we (and the iPad) mean by *display order,* see **Figure 1.30**.

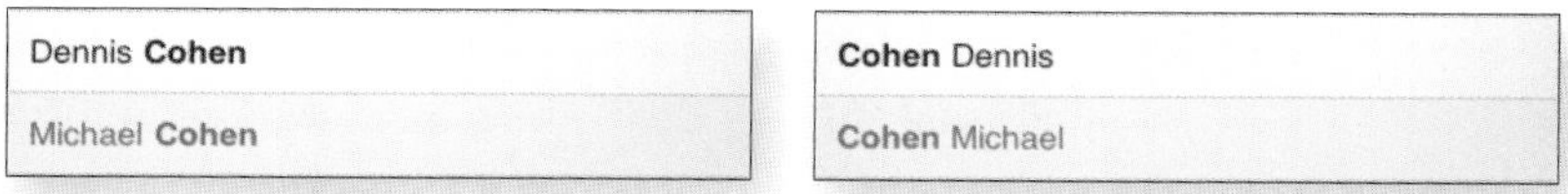

Figure 1.30 Two of this book's authors listed in the Contacts app in each of the available display orders.

7. Press the Home button to close the page and return to the home screen.
8. Open the Contacts app.

    Your contacts appear in the sort order that you specified, with each contact appearing in the specified display order.

Set a calendar

Your iPad's Calendar app can display events from a lot of calendars. You can have a Home calendar, a Work calendar, a Holidays calendar, a Bills Due calendar, and more. Adding to the fun, the Calendar app can display calendars from several sources: those that come from your computer, those that you sync with MobileMe, those to which you subscribe, and so on. But on what calendar do the events that you create on your iPad go?

That's where the default calendar comes in. You can (and should) specify a default calendar so that when you create an event, you don't have to worry about what calendar will receive it. Unless you specify otherwise, newly created events end up in your default calendar.

 Although you can subscribe to and sync with a plethora of calendars from various sources, the iPad doesn't give you a way to create new calendars directly on it.

Setting a default calendar:

1. Tap Settings > Mail, Contacts, Calendars.
2. Swipe down the resulting Mail, Contacts, Calendars screen until you see the Calendars section, way down at the bottom.
3. Tap Default Calendar.

    The Default Calendar screen appears, listing some or all of the calendars displayed by your iPad's Calendar app (**Figure 1.31**).

 We say *some or all* because the iPad shows only the calendars to which you can add events. Subscribed calendars, for example, can't be altered, so they don't show up in the Default Calendar screen.

Figure 1.31 You can pick a default calendar from those on your iPad.

4. Tap the calendar that you want to use as your default.

   The next time you create an event in the Calendar app, the event is placed in your default calendar—that is, unless you choose another calendar for the event when you create it. For an example, see the Calendar setting in the Add Event dialog shown in **Figure 1.32** on the next page.

 If your iPad isn't running iOS 4 or later, you can't move an event to a different calendar on your iPad after you create it, so remember to double-check the Calendar setting in the Add Event dialog before you save the event.

Figure 1.32 You can assign an event to a nondefault calendar when you create it.

Hear calendar alerts

The calendar-alert settings are relatively minor—unless you miss an appointment or invitation because your iPad happened to be asleep at the critical time. The iPad has two sound-related settings that are associated with your calendar: Calendar Alerts and New Invitation Alerts. When these settings are turned on, your iPad makes a sound when a calendar event is set and when one of your calendars receives a new invitation.

 Invitations go directly to one of your calendars in certain types of mail accounts, such as accounts on Microsoft Exchange servers.

Turning on alert sounds:

1. Tap Settings > General > Sounds to open the Sounds screen.

2. If the Calendar Alerts switch is off, tap it to turn it on.

3. In the Settings pane on the left side of the screen, tap Mail, Contacts, Calendars.

4. Swipe down the resulting Mail, Contacts, Calendars screen until you see the Calendars section at the bottom.

5. If the New Invitation Alerts switch is off, tap it to turn it on.

   From now on, whenever one of your calendars needs to alert you about an event or an invitation, your iPad makes a sound.

Use Time Zone Support

The Time Zone Support feature on your iPad seems to cause no end of confusion, if the number of posts on the Internet about it means anything. We'll try to clarify.

Your iPad has a Time Zone setting in its Date & Time settings page. When you set your device for a different time zone, all the events in your calendars shift their times to accommodate that change.

Your iPad also has a Time Zone Support setting in its Calendar settings page. When you turn that setting on and set a time zone, all the events in your calendars shift their times to match *that* time zone, regardless of what your iPad's Date & Time time-zone setting happens to be.

We'll wait until your head stops spinning. Feeling better? Good. Now, here's how you might use that feature.

Suppose that you're traveling from Los Angeles to Boston via a connecting flight in Chicago. Your flight from Los Angeles leaves at 8 a.m. Pacific time. According to your flight itinerary, your connecting flight to Boston from Chicago leaves at 2 p.m. Central time. Here's what you do:

1. On your calendar, enter two events: your takeoff at 8 a.m. and your connecting flight at 2 p.m. (Yes, even though the second flight takes off at noon Los Angeles time, you don't need to do the math to figure that out.)

2. Turn on Time Zone Support, and set the time zone for Los Angeles.

3. Go catch your 8 a.m. flight out of LA.

4. When you land in Chicago, your iPad's clock is 2 hours off, so set your iPad's Date & Time time-zone setting to Chicago time.

   Now your iPad's clock is correct, but because you turned on Time Zone Support, the events in your calendars don't move. The departure time for your connecting flight is still 2 p.m. on your calendar, which is where you want it to be.

That's all there is to it. For everything to work properly, of course, your iPad needs to have its Date & Time setting properly set for Los Angeles when you start.

Setting the Date & Time time-zone setting:

1. Tap Settings > General to open the General settings screen.

2. Tap Date & Time to open the Date & Time screen.

3. Tap Time Zone to open the Time Zone screen.

4. In the search field at the top of the screen, begin to type the name of the closest big city in your time zone.

   A list of city names appears below the search field (**Figure 1.33**).

Figure 1.33 Pick a city in your time zone.

5. Tap a city that's in the same time zone as you.

6. If necessary, tap the Date & Time arrow at the top of the screen to return to the Date & Time page, and set your iPad's clock to the current time.

Setting Time Zone Support:

1. Tap Settings > Mail, Contacts, Calendars.

2. Swipe down the resulting Mail, Contacts, Calendars screen until you see the Calendars section; then tap Time Zone Support.

   The Time Zone Support screen appears.

3. If the Time Zone Support switch is off, tap it to turn it on.

4. Tap Time Zone to open the Time Zone screen.

5. In the search field at the top of the screen, begin to type the name of a city in the time zone that you want.

   A list of cities appears below the search field.

6. Tap the city that's in the time zone you want.

   You're done!

Turn off Time Zone Support when you're not using it. If it's on and set for a different time zone from the one you're in, your calendar events may appear at the wrong times the next time you sync your iPad.

Get directions

This feature is a fun one to use and a real time-saver. In your Contacts app, every street address for each contact is linked to your iPad's Maps app. Here's how to use it.

Seeing a contact's address with the Maps app:

1. Open the Contacts app.

2. Choose a contact who has a street address.

3. Tap that street address.

   The Maps app opens, showing the location of the address you just tapped.

This feature is particularly helpful when you're traveling. Suppose that you're in a hotel, and you have a meeting with a client at her office. It's a simple matter to get directions from your hotel to your client's office. Read on.

Getting directions from your location to a contact's address:

1. In the Contacts app, tap a contact's address.

    The Maps app opens, with your contact's address pinpointed on the map and the address displayed in the search field in the top-right corner.

2. Tap the Directions button in the top-left corner of the map.

    The contact's address appears in the rightmost field at the top of the map, which is the Destination field.

 3. Tap the Current Location icon.

    Current Location appears in the Start field, which is just to the right of the Destination field. Also appearing is a list of instructions that tell you how to get from where you are to where your contact is and the route marked on the map. You can choose among walking, driving, and public-transit instructions at the bottom of the map.

2

Working and Playing in the iPad

Conventional wisdom has it that the iPad is just a media-consumption device.

Far be it from us to argue with conventional wisdom. Nonetheless, in this chapter we show you how to do more with your iPad than just lean back and consume books, videos, Web pages, and music.

Turn the page and follow along as we demonstrate how to get stuff onto your iPad so you can do things with that stuff. Then we show you how to make stuff with the stuff you have on your iPad—stuff like party invitations, instructional flash cards, fine food, and vacation plans.

Why do we do all this? Because the iPad is more than just a media-consumption device. That's our unconventional wisdom.

File Management Project

Difficulty level: Easy

Software needed: iTunes 9.1 or later, Apple's Pages for iPad or Mac ($9.99 from the App Store), DataViz's Documents To Go – Office Suite ($9.99 from the App Store), Dropbox for your iPad and for your computer (free)

Additional hardware: None

Your iPad is fabulous, and you take it just about everywhere you go. But sometimes, getting files to and from your iPad and your computer or other device can be a little tricky. This project talks about moving files back and forth among your iPad, your computer, and the 'net, using Pages for the iPad and the Mac, Documents To Go – Office Suite, and the free version of Dropbox for your iPad and your computer.

Email files to yourself

Probably the easiest way to move a file between your computer and your iPad (or smartphone) is to send the file as an email attachment. You create the email on your computer and attach the file you want to be able to use on your iPad. Then you send the file to yourself, using an address in a mail account that you can access on your iPad. Mail on the iPad can preview PDFs; iWork files; and Microsoft Word, Excel, and PowerPoint files.

There's a catch, however: The iPad can preview only certain file formats (see the nearby sidebar "Mail Attachments That the iPad Can Preview"). You can view these files in Mail—or see them or hear them—but you can't edit them in Mail. If you want to modify or edit these files, you need another app. We describe some of the ways to edit files later in this project.

If you do have an app that can edit a file type, Mail may know about it and ask whether you want to open the attached file in that app— another task that we cover later in this project.

If Mail or the iPad doesn't support the format of an attached file, you see the name of the attached file in the body of the email, but you can't open it on your iPad. You may be able to open it on a computer, however.

Emailing yourself a file to preview on the iPad:

1. Using a computer instead of your iPad, address an email to yourself.

   Be sure to use an address in an email account that you've set up on your iPad (see the **Mail Management Project** in Chapter 1). Fill out the Subject line, too, so that your Internet service provider won't think it's spam.

2. Attach a file that you want to read on your iPad.

   We've included a list of files that the iPad and Mail can read natively in the "Mail Attachments That the iPad Can Read" sidebar. For the purposes of this task, we're going to assume that you've attached a Microsoft Word .doc file.

3. Send the email to yourself.

In the next task, we're going to talk about previewing an attached Word file in Mail on your iPad. The process for previewing other kinds of files on your iPad is very similar.

Mail Attachments That the iPad Can Preview

By *preview*, Apple means that you can view, read, or play the following file formats within Mail. You can't edit previewed files, though you may be able to use the Copy command.

- **Adobe Acrobat and Preview (Mac) files:** .pdf
- **ASCII/text files:** .txt
- **Audio files:** .aac, .aiff, .mp3, and .wav (all of which you can play in the Mail app)
- **Image files:** .gif, .jpg, and .tiff (displayed as inline images in Mail)
- **iWork files:** .key (Keynote), .numbers (Numbers), and .pages (Pages)
- **Microsoft Office files:** .doc and .docx (Word), .ppt and .pptx (PowerPoint), and .xls and .xlsx (Excel)
- **Rich-text files:** .rtf
- **vCard files:** .vcf (which you can import into Contacts; see the **Mail Management Project** in Chapter 1)
- **Web pages:** .htm and .html

Previewing a Microsoft Word file in Mail:

1. Open an email with an attached Word file (see the preceding task).

   You see an icon in the body of the email like the one shown in **Figure 2.1**. (If you've installed another app that can read Word files on the iPad, you might see that app's icon instead.)

Figure 2.1 Icon for an attached Microsoft Word file in Mail.

2. Tap the icon.

   A preview screen opens (**Figure 2.2**).

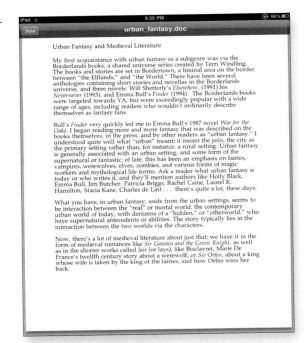

Figure 2.2 Mail showing a preview of a Word document.

3. Read the document.

   If the attached file is a multiple-page document, you can swipe up or down to page through the file. You can copy and paste from the previewed document, though you can't edit it.

4. When you finish reading, tap the blue Done button in the top-left corner.

   The preview screen disappears, and the email you began with is revealed.

   If you have Pages for iPad installed, you can open and edit Word or Pages files that have been sent as email attachments. The procedure for opening and importing an email attachment in Pages is very similar to the procedure for opening email attachments in other apps. In the following task, we show you how to open an emailed Word file in Pages for iPad.

Opening an emailed Word file in Pages:

1. View an email with an attached Word file in Mail (refer to "Emailing yourself a file to preview on the iPad" earlier in this project).

2. Tap the Word attachment.

   Because you have Pages for iPad installed on your iPad, the attached file's icon displays the Pages icon (**Figure 2.3**).

Figure 2.3 Icon of an attached Word file when Pages for iPad is installed.

3. Tap the icon.

   You see the attached file in the Mail preview screen (**Figure 2.4**).

Figure 2.4 Mail preview of a Word file, showing an Open in Pages button.

This screen looks very similar to the one shown in Figure 2.2 earlier in this project, but there's one small difference: Figure 2.4 has a navigation button in the top-right corner that takes you to an app that can read Word files. That button may say Open in Pages, or it may list another app on your iPad that can read and edit Word files.

4. Tap the Open in Pages button to open the attached file in the Pages app.

Mail quits while Pages launches, displaying an import screen with a progress bar (**Figure 2.5**).

Figure 2.5 Pages import screen for a Word file.

 Depending on the nature of the document you imported into Pages from Word, you may see a dialog alerting you to an incompatible font (**Figure 2.6**). Read the message and then tap the blue Done button in the top-right corner.

Chapter 2: Working and Playing in the iPad

Figure 2.6 A Pages import warning about a font substitution.

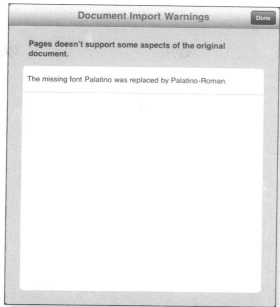

When the import finishes, you see your Word file in Pages' My Documents screen, ready to read, edit, or save (**Figure 2.7**).

Figure 2.7 Your Word file in Pages.

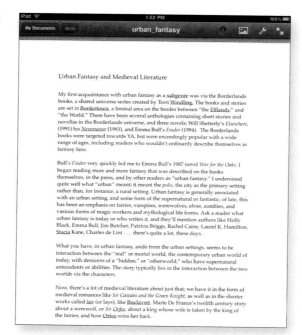

5. When you finish viewing the document, press the iPad's Home button to exit Pages and return to the home screen.

Pages for iPad Compatibility

Pages can import and export various images and file formats:

- **Imports:** Pages '09 for Mac, Word (Office Open XML and Office 97 or later .doc)
- **Exports:** Pages '09 for Mac, Word 97 or later, PDF
- **Inserts:** Images from your iPad's Photos app

Apple lists the fonts supported by Pages for iPad and discusses formatting imported documents at http://support.apple.com/kb/HT4065.

note Importing a Pages '09, Keynote, or Numbers file sent as an email attachment is very similar to importing a Word file; the only difference is the application you use to import the file. Numbers imports Excel files, and Keynote can import most PowerPoint slides.

Transfer a file from a computer

Email isn't the only way to transfer files to your iPad. You can use iTunes to copy files to your computer from your iPad and from your iPad to your computer. When your iPad is connected to your computer via a USB port and the special 30-pin iPad cable, the iTunes Apps tab contains a File Sharing section.

Transferring a Word file to your iPad via iTunes:

1. Connect your iPad to your computer.
2. Launch iTunes, if it doesn't open automatically.
3. Select your iPad in the iTunes Source list.
4. In the main iTunes window, click the iPad's Apps tab.
5. Scroll down until you see the File Sharing section (**Figure 2.8**).

Chapter 2: Working and Playing in the iPad   **61**

> **note** This section may look different if you have different apps installed on your iPad or if you use Windows instead of Mac OS X.

Figure 2.8 File-sharing options.

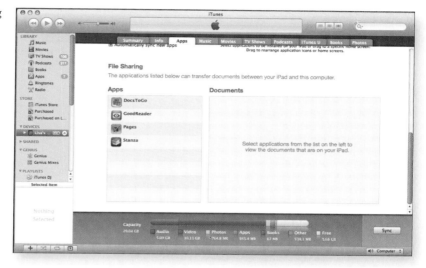

6. Select Pages in the Apps list on the left side of the tab.

   A Pages Documents list appears on the right side of the tab (**Figure 2.9**).

Figure 2.9 Pages Documents list.

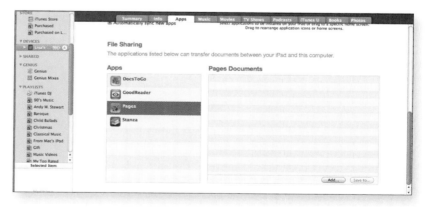

7. Click the Add button in the bottom-right corner of the tab.

   You see a familiar Open dialog for Windows or for Mac.

8. Navigate to and select the file you want to transfer to your iPad; then click the Choose button.

   Your selected document appears in the Pages Documents list (**Figure 2.10**).

> **On a Mac, instead of completing steps 7 and 8, you can drag a file from a Finder window to the Pages Documents list.**

Figure 2.10 Pages Documents list with a new file.

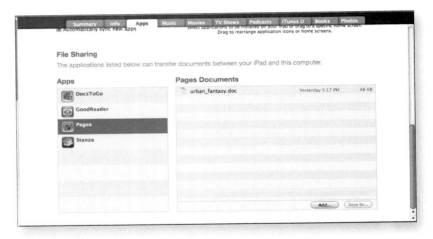

When the file is in Pages for iPad, you can modify it at will, including changing the formatting and editing or adding to it, as we show you in the following section.

Import files into the iPad

You've transferred a file to Pages for iPad via iTunes. Now you want to import the file into Pages for iPad. The following tasks assume that you're working with a Word file and Pages for iPad, but the process is similar for importing Keynote or PowerPoint files into Keynote for iPad, or Excel or Numbers files into Numbers for iPad.

Importing a Word file into Pages for iPad:

1. Launch the Pages for iPad app on your iPad.
2. If you aren't already viewing the My Documents screen, tap the My Documents button in the top-left corner of the Pages screen.

Chapter 2: Working and Playing in the iPad 63

 3. Tap the folder icon in the top-right corner of the My Documents screen.

You see the Import Document screen, listing the documents you've transferred to Pages for iPad. **Figure 2.11** shows an example; your list of documents will be different.

Figure 2.11 Import Document screen.

4. Select a document to import.

An import screen with a progress bar appears briefly (refer to Figure 2.5 earlier in this project).

 Depending on the formatting and fonts used in your original document, you may see an alert dialog like the one shown in Figure 2.6 earlier in this project. If you do see this dialog, read the message and then click the blue Done button.

When the import is finished, you see your document in Pages for iPad. You can find the document later by tapping the Pages icon and then tapping the My Documents button if the document you want isn't currently open in Pages for iPad.

 Use the same process to import a Pages '09 file into Pages for iPad.

Export files from the iPad

You can export files from Pages for iPad as Word files, Pages '09 files, or PDF files. The process is the same, even if the file formats are different.

Exporting a file from Pages to Word:

1. Launch the Pages app on your iPad.
2. Tap the My Documents button if you're already viewing a document.
3. In the My Documents screen, swipe left or right to page through your documents until you find the one you want to export.
4. Tap the curved-arrow Export or Share button at the bottom of the document.

   A floating window like the one shown in **Figure 2.12** offers you three choices: Send via Mail, Share via iWork.com, and Export.

Figure 2.12 Pages for iPad options for sharing and export.

5. Tap Export.

   The Export Document screen opens (**Figure 2.13**).

Figure 2.13 Export Document screen.

6. Tap the Word icon to export the document in Word format.

   You see a progress bar as the file is converted and exported (**Figure 2.14**).

Figure 2.14 Exporting a file in Word format.

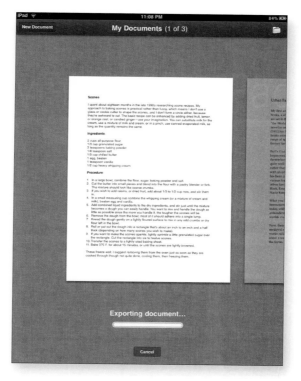

You've exported the file, converting it from Pages for iPad format to Word format, but you still need to get it on your computer. You could email it to yourself (or to someone else) from your iPad, as we show you earlier in this project, but you can also use iTunes to transfer it, as we show you next.

Transferring a Word file to your computer via iTunes:

1. Connect your iPad to your computer.
2. Launch iTunes, if it doesn't open automatically.
3. Select your iPad in the iTunes Source list.
4. In the main iTunes window, click the iPad's Apps tab.

5. Scroll down to the File Sharing section (refer to Figure 2.8 earlier in this project).

6. Select Pages in the Apps list on the left side of the tab.

   You see all documents that you exported earlier in the Pages Documents list (**Figure 2.15**).

Figure 2.15 Pages Documents list showing an exported file.

7. Select the document you want.

8. Click the Save to button in the bottom-right corner of the tab.

   The familiar Save dialog opens.

9. Specify where you want to save that file on your computer.

10. Click Choose.

    Your iPad transfers the file to your computer, where you can open the exported file and continue working on it.

There are other ways to transfer documents between your iPad and a computer besides using email or iTunes. For one, Apple offers the iWork.com site. When this book went to press, iWork was still in beta and was free, but that situation may change.

Another way to manage and transfer files is to use Dropbox, which supports sharing, syncing, transferring, and backing up files. We cover that method in the following task.

Use Dropbox

Dropbox allows you to share data between computers and other devices by storing your files on Dropbox's servers. A free account gives you 2 GB of file storage. You can access any files that you upload from any device that has Dropbox installed or from the Dropbox Web site.

To use this service, register for a free account at www.dropbox.com, and download the software for your operating system (Mac, Windows, or Linux); also, download the free Dropbox for iPad app from the App Store.

You can share any file in your Dropbox by dragging a file into your Public folder in Dropbox on your iPad, and you can sync files that you have marked as Favorites in Dropbox on your iPad. Those files will sync automatically with their twins on the Dropbox server, so you can edit them on any device and always have access to the current versions.

When you have a file on Dropbox (you can upload files via the Web site, or drag and drop files from your computer to Dropbox), you can mark it as a favorite on your iPad, as we show you in the following task. We're going to use a Word file for this task, but you can favorite and sync any file that you can edit on an iPad.

Favoriting a file in Dropbox on your iPad:

1. Launch the Dropbox app on your iPad.

    You should see a screen titled My Dropbox (**Figure 2.16**). If you don't, tap the Dropbox icon on the far-left side of the Dropbox screen.

Figure 2.16
My Dropbox screen.

2. Tap a file in your My Documents list.

   The file opens in Dropbox's preview screen (**Figure 2.17**).

Figure 2.17 Dropbox preview of a Word file.

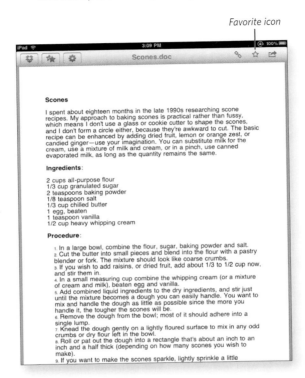

Favorite icon

3. Tap the single star icon at the right end of the screen's toolbar.

   The star turns dark to indicate that you've marked the file as a favorite.

   Now if you edit the file in another app or on your computer by opening the file directly in Dropbox, the file will sync.

You can force the file on your iPad to update when you change it on another device in Dropbox by viewing the document in Dropbox and tapping the double-star icon at the left end of the toolbar. You see your favorites list, which has a button that allows you to update all files or selected files.

Dropbox works in tandem with other iPad apps, including Documents To Go, which we cover in the next section. You can transfer a file you want to edit via Dropbox and then edit it (or convert it) in Documents To Go.

Use Documents To Go – Office Suite

Documents To Go lets you view, edit, create, and manage Word and Excel files on your iPad. It also allows you to view other file types supported by the iPad, including iWork, PDF, and image files. The Office Suite version lets you manage local files on your iPad transferred via iTunes, Dropbox (see the preceding section), or email.

 When you open Documents To Go on your iPad for the first time, a Getting Started wizard opens. It's a good idea to read through it for a quick introduction, because the software has some surprisingly robust editing and file-creation capabilities that are easy to miss.

Importing an emailed Word file into Documents To Go for editing:

1. Locate an email with a Word attachment.

2. Tap the Word icon.

    You see a preview of the file (**Figure 2.18**).

Figure 2.18 Mail preview screen.

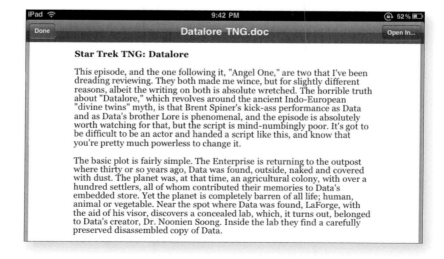

3. Tap the Open In button in the top-right corner.

 Depending on the number of apps you've installed, your button may simply say Open instead.

You see a floating window similar to the one shown in **Figure 2.19**.

Figure 2.19 Document options.

4. Tap DocsToGo.

   You see the Documents To Go splash screen briefly while the app loads the Word file. Then your document opens in the Documents To Go preview screen (**Figure 2.20**).

Figure 2.20 Documents To Go preview screen.

Chapter 2: Working and Playing in the iPad 71

5. To save the file locally, tap the Save button in the bottom-left corner (resembling an old-style floppy disk).

   A palette like the one shown in **Figure 2.21** pops up from the button.

Figure 2.21 Save palette.

Save button

6. Tap the icon with the curved arrow on the right side of the palette.

   A Save As screen opens (**Figure 2.22**).

Figure 2.22 Save As screen.

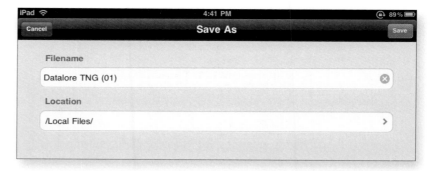

7. Change the file name, if you want.

   Documents To Go automatically names the file by appending a number in parentheses to the original name. You can keep that name or change it by tapping the x at the right end of the Filename field and entering a new one.

8. Tap the blue Save button in the top-right corner of the screen.

   Now you can edit the file in Documents To Go. Remember to save it regularly.

 In the bottom-left corner of the Documents To Go preview screen (refer to Figure 2.20) is a Mail button, marked by an envelope icon. Tap this icon to launch Mail, which presents an blank email with the edited file attached, ready for you to address and send.

File Transfer with Phone Disk

Phone Disk is a program for Intel-based Macs that mounts iPads and other iOS devices on the Desktop so you can move files between your Mac's Finder and your iPad by dragging and dropping. The software automatically detects when you've connected an iPad or other iOS device to your Mac, and it gives you access to your iPad's Media and Document Sharing folders.

You can download Phone Disk at www.macroplant.com/phonedisk. The software is free until December 1, 2010.

iPad Chef Project

Difficulty level: Easy

Software needed: Epicurious app (free), BigOven Lite app (free), Apple's Pages for iPad ($9.99)

Additional hardware: Kitchen and camera (optional)

The iPad, in part because of its form factor and in part because the display is so gorgeous, was almost immediately put to work in the service of good food and great cooks. Cooking sites released image-laden apps that allowed users to browse, search, bookmark, and drool over fabulous recipes. The larger food and cookery sites realized right away that cooks want an easy way to generate printable, shareable recipes and shopping lists. They also realized that cooks were going to be following recipes on their iPads in the kitchen while they cooked.

This project is about finding recipes with the Epicurious and BigOven Lite apps, favoriting the recipes you want to try, emailing recipes that you want to share, and creating your own recipe scrapbook in Pages for iPad.

Find recipes with Epicurious

Long before the iPad, there were cookbooks; glossy cooking and food magazines, many with equally glossy Web sites; and large community sites for cooks and lovers of food of all sorts. There still are. With so many resources available, searching for just the right recipe can be confusing.

Fortunately, iPad apps are available to save you time and effort. Epicurious.com, the Web home of *Epicurious* magazine, recently released a version of its iPhone app for the iPad. This free app (available at the App Store) makes it easy to browse or search thousands of recipes provided by *Gourmet* and *Bon Appetit* magazines, professional chefs, popular cookbooks, and famous restaurants. Here's how.

Using Epicurious to browse recipes:

1. Tap the Epicurious icon on your iPad to launch the app.

   The Epicurious splash screen appears very briefly, followed by the home screen, with its control panel open. On startup, the control panel displays the Featured category, as shown in **Figure 2.23**. (The categories change based on season, so your control panel will be different from the figure.)

Figure 2.23 Epicurious home screen.

2. Tap one of the categories in the control panel to browse recipes by type.

   You see a screen that looks very much like a page in a printed cookbook, with navigational tabs along the right side (**Figure 2.24**).

3. Tap a recipe to see that recipe in its own screen, with the Ingredients floating window open (**Figure 2.25**).

Figure 2.24 Category screen.

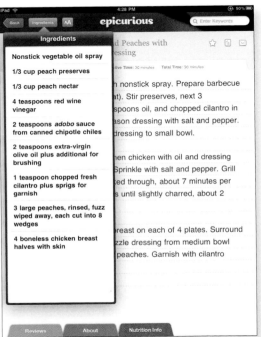

Figure 2.25 Recipe screen with Ingredients window.

4. Tap the recipe to make the Ingredients window disappear and see information about preparing the dish (**Figure 2.26**).

Chapter 2: Working and Playing in the iPad   75

Figure 2.26
Recipe screen.

5. For purposes of this project, tap the red Back button in the top-left corner of the recipe screen to return to the preceding category screen.

 A display bug in version 2.1.1 of Epicurious may prevent the red Back button from appearing. If you're viewing a single-recipe screen and don't see this button, rotate your iPad to change its orientation. The button will appear.

Navigating Epicurious

Thousands of recipes are available via Epicurious, and more are added all the time. Navigating this sea of recipes can be tricky, so here are a few hints:

- Tap a tab on the right side of a category screen to see the relevant view: Photo (recipes with images), Rating (recipe ratings by other cooks), A-Z (recipes in the category listed alphabetically), or Newest (most recently added recipes in the category).

- The colored navigation tabs at the bottom of every recipe screen provide more information. Reviews displays comments by other cooks who have tried the recipe; About describes the source of the recipe; and Nutrition Info lists the calories, grams of fat, and grams of fiber in a serving. Some recipes offer other tabs that provide additional information.

- Each recipe screen has an orange bookmark (refer to Figure 2.26). You can slide this bookmark up or down to mark your spot in the recipe, which makes it easy to keep track while you work in the kitchen.

Searching for recipes in Epicurious:

1. If a category screen isn't already open on your iPad, complete steps 1–2 of "Using Epicurious to browse recipes."

2. Tap the gray Control Panel button in the top-left corner of the screen.

    You return to the home screen, where you find the control panel open (refer to Figure 2.23 earlier in this project).

3. Tap the Search button at the top of the control panel to perform more advanced searches.

    This feature lets you search for a specific type of food or drink, the main ingredient in a recipe, or a cuisine; it also lets you search by dietary restriction or special occasion.

 You can also search via the Search field in the top-right corner of each recipe screen, but that field limits you to a keyword search.

Chapter 2: Working and Playing in the iPad   **77**

When you find a recipe you like by browsing or searching, you can favorite the recipe, much as you might bookmark a Web site, so that you can find it easily later. In the next task, we show you how.

Favoriting Epicurious recipes:

1. Tap the Epicurious icon to open the app, if it isn't already open.

2. Locate a recipe that you want to be able to find later (see the preceding two tasks).

   When you're looking at that recipe's page, you'll find a white star to the right of the recipe title (refer to Figure 2.26).

3. Tap the star to favorite the recipe.

   The star turns blue to show that the recipe has been added to your favorites list.

If you want to remove a recipe from your favorites list, tap its blue star. The recipe is deleted from the list, and its star reverts to white.

Displaying your Epicurious favorites:

1. Launch Epicurious, if it isn't already open.

2. If the control panel doesn't appear on the home page, tap the gray Control Panel button in the top-left corner.

3. Tap the Favorites button at the top of the control panel.

   The My Favorites screen opens (**Figure 2.27** on the next page). (Yours will be different from ours.)

   Each recipe in the My Favorites list has three icons to the right of its title, as shown in **Figure 2.28** on the next page. These icons all work the same way wherever they occur in Epicurious.

Figure 2.27
My Favorites screen.

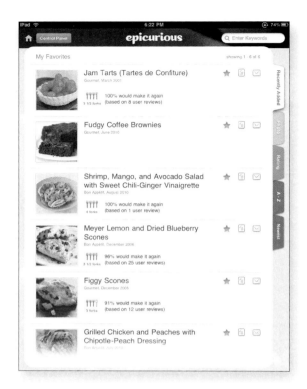

Figure 2.28
My Favorites recipe icons.

Viewing and emailing a recipe's shopping list in Epicurious:

1. Display your My Favorites screen (see the preceding task).

2. Tap the Shopping List icon to the right of a recipe's name (refer to Figure 2.28) to generate a shopping list for that recipe.

   If you tap the Shopping List icon for other recipes in your list of My Favorites, those shopping lists are combined with the first.

3. Display the shopping list by tapping the Control Panel button in the top-left corner to open the Control Panel screen and then tapping My Shopping List at the bottom of that screen.

   You see a nicely laid out My Combined Shopping List screen (**Figure 2.29**).

Figure 2.29
My Combined Shopping List screen.

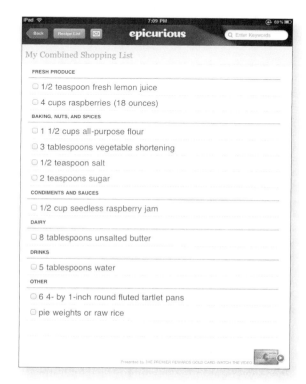

4. To email yourself a copy of a recipe, tap the red Back button in the top-left corner of the My Combined Shopping List screen to return to your My Favorites screen; then tap the email icon to the right of the recipe you want to send.

   Epicurious creates a blank email ready for you to add addresses, with the Subject line listing the recipe's title and the body of the email containing the recipe (**Figure 2.30** on the next page).

Figure 2.30 Epicurious recipe in an email, ready to send.

5. Add an email address and a brief message at the top of the email body, before the recipe.

   For this task, address the email to yourself.

6. Tap the Send button.

   In a minute or two, if you check your email, you should see a nicely formatted recipe in your Inbox.

Find recipes with BigOven Lite

BigOven began as a Web site for people who love to cook but who face the never-ending question "What do I make for dinner tonight?" on a regular basis. The BigOven Web site and apps are designed to help people plan meals and exchange recipes.

Two apps are available for the iPad: BigOven Pro ($9.99) and BigOven Lite (free). The paid app allows you to create grocery shopping lists that sync across devices, email grocery lists, and look up terms in a glossary. It's also ad-free. In the following tasks, you'll be using the free app, BigOven Lite, though the steps work for both versions.

Searching for recipes in BigOven:

1. Tap the BigOven Lite icon to launch the app.

   You see the splash screen, which is different each time you launch the app, though the toolbar is always visible (**Figure 2.31**).

Chapter 2: Working and Playing in the iPad   **81**

Figure 2.31 BigOven Lite splash screen.

2. Tap the BigOven button in the top-left corner.

    You see the BigOven navigation form (**Figure 2.32**), which displays categories you can search.

Figure 2.32 Navigation form.

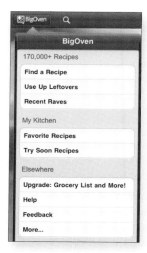

3. Tap Find a Recipe.

   A search form opens (**Figure 2.33**).

Figure 2.33 Search form.

You can perform a variety of searches, including restricting your search to the title of a recipe or keywords. The basic functions of all the search types are the same, however, so we'll go through one search step by step and leave the others for you to explore on your own.

4. Tap the Leftovers button at the top of the search form.

   You see a search window similar to the one shown in **Figure 2.34**.

5. Enter up to three "leftover" ingredients to search for recipes that use those ingredients (**Figure 2.35**).

Figure 2.34 Leftovers search form.

Figure 2.35 Leftovers search form showing search terms.

Chapter 2: Working and Playing in the iPad   **83**

6. Tap the large red Search BigOven button to see a list of recipes that match the ingredients you chose.

    You'll see a list similar to the one shown in **Figure 2.36**.

Figure 2.36 Leftovers search recipe results.

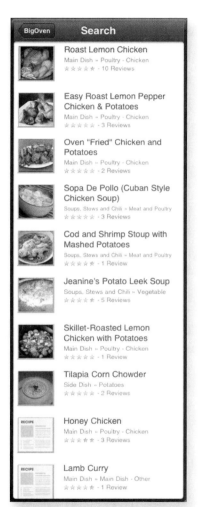

7. Tap a recipe in the list of search results.

    The app displays the recipe in easy-to-read form (**Figure 2.37** on the next page).

Figure 2.37
Recipe screen.

 Click the Prepare button in the toolbar to view the recipe without ads or extraneous data so you can follow the instructions easily when you're preparing the dish.

8. If you want to return to the search results from a recipe screen, tap the BigOven button in the top-left corner.

When you've found a recipe you like, it's useful to mark it as a favorite so that you can find it again easily, as we show you in the following tasks.

Marking favorite recipes in BigOven:

1. Tap BigOven Lite's icon to open the app, if it isn't already open.
2. Find a recipe you like (see the preceding task).

3. Tap the Export button at the far-right end of the toolbar (refer to Figure 2.37 earlier in this project).

   You see the options shown in **Figure 2.38**.

Figure 2.38
Export options.

4. Tap Add to Favorites.

   BigOven Lite adds the recipe to your favorites list.

Viewing favorite recipes in BigOven:

1. Tap the BigOven button in the top-right corner of the screen.

   You see the navigation form (refer to Figure 2.32 earlier in this project).

2. In the My Kitchen section, tap Favorite Recipes to see your favorites (**Figure 2.39**).

Figure 2.39
Favorites list.

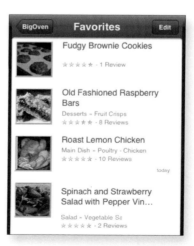

After you find and favorite recipes, you can email them to yourself or to a friend, as you see in the next task.

Emailing a recipe from BigOven:

1. Launch BigOven Lite, if it isn't already running.

2. Find a recipe by searching for it (see "Searching for recipes in BigOven" earlier in this project), or tap a recipe in your favorites list (see the preceding task).

3. Tap the Export button at the far-right end of the toolbar.

    You see the export options (refer to Figure 2.38).

4. Tap Email it.

    A formatted email message opens (**Figure 2.40**), with your default email address displayed in the From field.

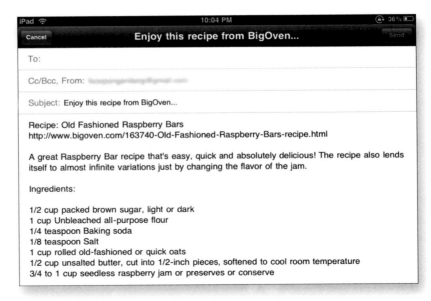

Figure 2.40 Recipe ready to email.

5. Enter the recipient's address, add a short message, and tap Send to send the email.

Now that you know how to find, favorite, and email recipes, the next step is creating a personal recipe collection on your iPad that you can add to at will. In the following tasks, you'll use Pages for iPad to create a recipe scrapbook.

Create a recipe scrapbook

Creating a recipe scrapbook is a great way not only to preserve your own recipes, but also to keep recipes that you've collected from the Internet or from friends who emailed their recipes to you. You can create an attractive recipe scrapbook quite easily in Pages for iPad. The process involves four steps:

1. Get the recipe text into Pages.
2. Prepare the dish (which we'll let you do on your own).
3. Take a picture of the finished dish.
4. Import and position the image in the Pages recipe.

You can enter a recipe directly in Pages or email it to yourself as a Microsoft Word file, an .rtf file, or a Pages file attachment and then import the file into Pages. For step-by-step instructions, see the **File Management Project** earlier in this chapter.

In the following tasks, we show you how to start your recipe scrapbook by importing an emailed recipe into Pages for iPad and then formatting it, adding artwork. You'll be working with the recipe for Old Fashioned Raspberry Bars from BigOven, which you can find in the BigOven Lite app by searching for it by title (see "Searching for recipes in BigOven" earlier in this project).

Importing an emailed recipe into Pages:

1. Open a recipe that you received by email (see "Emailing a recipe from BigOven" earlier in this project).

    The recipe will have basic formatting even in email form (**Figure 2.41**).

Figure 2.41 Recipe emailed from BigOven Lite.

Tap and hold here to display selection buttons.

2. Tap and hold just to the right of where the text begins in the email body.

   You see the iPad Select and Select All buttons (**Figure 2.42**).

Figure 2.42 iPad selection buttons.

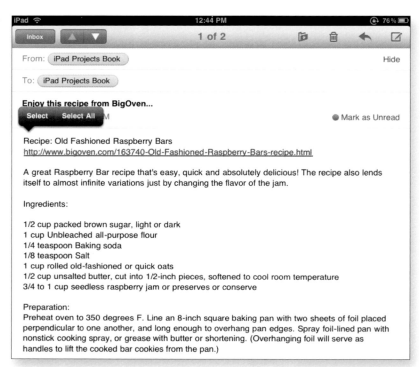

3. Tap Select All to select the entire body of the email—in this case, the recipe.

4. Tap the Copy button that appears.

   The button briefly flashes blue to indicate that the selected text has been copied to your iPad's internal clipboard, ready to paste.

   At this point, you could paste the selected text into the Notes app on your iPad, Documents To Go, or one of the many simple note apps available for iPad, but for this project, you use Pages because you want to include images.

5. Launch Pages on your iPad.

6. If you don't see the My Documents screen when Pages opens, tap the My Documents button in the top-left corner.

Chapter 2: Working and Playing in the iPad   89

7. Tap the New Document button in the top-left corner of the My Documents screen.

    The Choose a Template screen opens (**Figure 2.43**).

Figure 2.43 Choose a Template screen.

8. Tap Blank.

    A blank document opens.

The document is automatically named Blank. You can change that title if you want in the My Documents screen (tap the word *Blank* below the file's icon and then type a new name), but for convenience, we'll refer to this document as Blank.

9. Tap and hold in the body of the document until you see the iPad Paste and Copy Style buttons (**Figure 2.44**).

Figure 2.44 Paste and Copy Style buttons.

10. Tap Paste.

The text you copied from the email in step 4 appears in the Pages document (**Figure 2.45**).

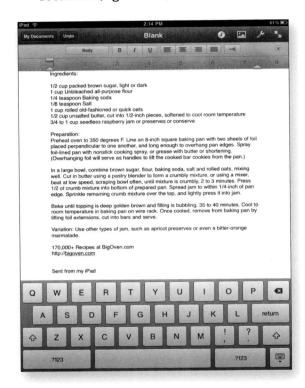

Figure 2.45
Text pasted from the iPad clipboard.

You'll use this Blank document as a staging document while you build the recipe scrapbook in the following task.

Creating a recipe file for your scrapbook:

1. Tap the Pages icon to open the app.

2. If Pages opens to a file instead of the My Documents window, tap the My Document button in the opening screen.

3. Tap the New Document button in the top-left corner of the My Documents screen.

4. Scroll down to tap the Recipe template.

   A document based on that template opens in Pages (**Figure 2.46**).

 Pages names the new file Recipe by default, and you can change that name, but we'll refer to the document as Recipe in this project.

Figure 2.46 Recipe template.

5. To change the title text from *Nona's Holiday Cookies,* triple-tap that text to select the entire line.

   A selection rectangle appears around the title.

6. Type new title text.

7. If you want to, tap and hold the text below the title to select and edit the cooking time and quantity of your recipe.

Next, you'll copy the ingredients for your recipe to the new scrapbook.

Copying recipe ingredients to the scrapbook:

1. Tap the My Documents button in the top-left corner of the recipe screen.

2. In the My Documents screen, tap to open the Blank document where you pasted the recipe (see "Importing an emailed recipe into Pages" earlier in this project).

3. Tap and hold to the left of where the ingredients list begins.

   Selection buttons appear (**Figure 2.47**).

Figure 2.47 Displaying selection buttons.

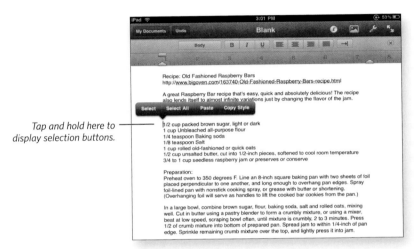

Tap and hold here to display selection buttons.

4. Tap Select, and drag the selection rectangle around the entire list of ingredients (**Figure 2.48**).

   As soon as you lift your finger off the screen, the Copy button appears.

Figure 2.48 Selecting the ingredients list.

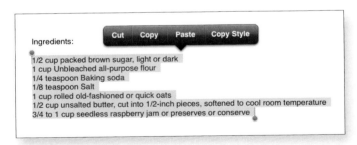

5. Tap Copy to copy the selected text to your iPad's clipboard.

6. Tap the My Documents button to open the My Documents screen.
7. Tap the Recipe document (see the preceding task) to open it.
8. Tap and hold to the left of the italic placeholder text that begins *Etiam sit amet est* to display selection buttons.
9. Tap Select.

   Pages selects the entire block of placeholder text and displays the Paste button (**Figure 2.49**).

Figure 2.49 Selecting placeholder text.

10. Tap Paste.

    The template's placeholder text is replaced by the ingredients for your recipe.

If you want your ingredients list to be in italics, select the ingredients and then tap the *I* button on the Pages toolbar.

11. Repeat steps 1–10, copying the procedure or preparation section of your recipe from the Blank staging document and pasting it in the Cooking Instructions section of the Recipe document (**Figure 2.50** on the next page).

Now all you need to do is replace the image from the template with the image for your recipe.

In the following tasks, you'll use an image from a Web site, though you can certainly use a photo you took yourself.

Figure 2.50
The Recipe document, with all the template text replaced by a new recipe.

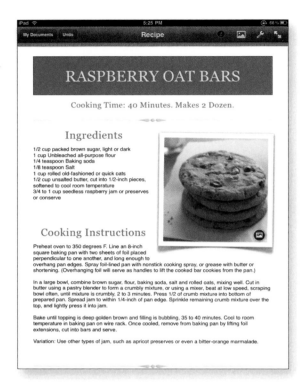

Customizing Recipes in Pages

You can extensively customize the documents that you make from Pages templates. Here are some ways you may want to customize your recipes:

- To change the font or color of text, select the text; tap the *I* button on the Pages toolbar; and then tap the Style and Text options in the form that appears. You can set the size of text, the color, and the font.

- To change the background color of the title in a recipe file, tap the colored background area to select it; tap the *I* button on the Pages toolbar; tap Style in the form that appears; and choose a new background color.

- It's a good idea to include a note about where a recipe originated or who created it. You can simply copy the URL from the email and paste it into the recipe, or you can type a short note about your source.

Importing an image into Pages:

1. Launch Safari on your iPad.

2. Go to a Web site that has suitable images.

   For this task, go to the Old Fashioned Raspberry Bars page at www.bigoven.com/163740-Old-Fashioned-Raspberry-Bars-recipe.html.

3. Tap and hold the image you want to use.

   Save Image and Copy buttons pop up (**Figure 2.51**).

Figure 2.51 Copying an image from a Web site.

4. Tap Save Image.

   The image is automatically downloaded and saved in the Saved Images album of your iPad's Photos app.

The next task replaces the placeholder image from the Recipe template with this image.

Inserting an image into Pages:

1. In Pages, open the Recipe document from "Creating a recipe file for your scrapbook" earlier in this project.

    If the document isn't currently open in Pages, tap the My Documents button; then tap the Recipe document's icon in the My Documents screen.

2. Tap the image icon in the bottom-right corner of the placeholder image.

   A Photo Albums floating window opens (**Figure 2.52** on the next page), listing the albums in your Photos app.

Figure 2.52 List of albums in Photos.

3. Tap Saved Photos.

   You see the images that you've saved to your iPad, including the one you saved from a Web page in the preceding task (**Figure 2.53**).

Figure 2.53 Saved Photos album, showing the newly saved image.

4. Tap the image you saved from the Web site.

   Your image replaces the placeholder image in the template (**Figure 2.54**).

Figure 2.54 The template image replaced by the correct image.

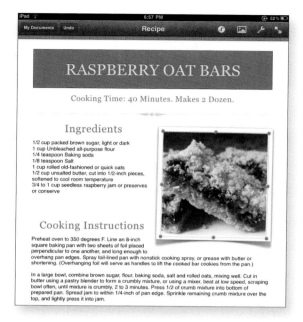

5. Adjust the size of the image, if you want, by tapping it and then dragging the selection rectangle.

6. Drag the image to a different place on the page, if you want.

Using an iPad in the Kitchen

Using an iPad in the kitchen can be truly labor-saving, but it can also be very risky. Electronic devices really aren't designed to spend time around bubbling liquids, for example. Here are a few tips to make cooking with an iPad easier and safer:

- **Stand it up.** Think about getting a stand for your iPad. An awful lot of people have found that a cookbook stand works perfectly for an iPad, though you may want to position it in a place that's somewhat sheltered.

- **Bag it up.** A gallon-size plastic food-storage bag (the kind with a zipper closure) makes a great protective cover for an iPad. Slip the iPad inside a clean dry bag, and seal it. The iPad will be a little less vulnerable to spills and splashes, and you can still use the touchscreen through the bag.

- **Don't let it sleep.** Consider changing the Auto-Lock setting (tap Settings > General) to keep your iPad from going to sleep and locking the screen at a crucial moment.

Party Project

Difficulty level: Easy

Software needed: iPad's Notes and Maps apps, Pages for iPad ($9.99)

Additional hardware: None

Yes, you can sit down at a table and plan a party with just a pencil and a pad of paper. But really, it's a lot of work keeping track of all of those slips of paper, and anyway, it's *such* a last-century thing to do.

You could avoid the paper shuffle by planning your party at the keyboard of your trusty computer, but…well, that seems like just another day at the office.

So how about this? Take advantage of your iPad's portability, connectivity, and lovely bright screen to tap out your party plans wherever you are and whenever the mood strikes you.

In this project, with iPad in hand, you'll coordinate your party's shopping and invitation lists; you'll create your invitations, complete with a map; and you'll track the RSVPs as they come rolling in, even while you're pushing your shopping cart down aisle 7B of the grocery store, loading it up with chips, cheese dip, and seven kinds of sparkling water.

Sound good? Well, then, let's get this party started!

Use Notes to make lists

Notes on the iPad is about as simple a text-based app as you can get. You use it to create notes, edit them, and search them. That's it. You can't set fonts or colors, but you really don't need them for jotting quick memos or simple lists.

What makes the preinstalled Notes app especially interesting for our purposes is that it's designed to work with your other apps and devices. You can email a note from within the Notes app, and you can sync your notes with iPhones and iPod touches, with Mail on Macs, and with Microsoft Outlook on Windows PCs. (You can read about syncing notes in the **Information Syncing Project** in Chapter 1.)

First, you'll use Notes to make a simple invitation list of the people to invite via email.

Chapter 2: Working and Playing in the iPad 99

Creating an invitation list in Notes:

1. Tap Notes to open the app.

   Notes automatically opens to the last note you viewed, so if you've used the app before, you may see a note you've already created instead of the blank one shown in **Figure 2.55**. To create a new blank note, tap the plus (+) button on the toolbar.

Figure 2.55 A blank note in the Notes app.

2. If you don't see a blinking cursor and the digital keyboard, tap the first blank line in the new note to make them appear.

3. Type **Invitation List** and then tap the Return key on the digital keyboard to move the cursor to the next line.

4. Add some names, tapping Return after each name so that each one is on a separate line.

   For this task, you'll use the names of film characters for privacy reasons, but pretend that Harris K. Telemacher, Bilbo Baggins,

Samwise Gamgee, James T. Kirk, Sam Spade, and Rick Blaine are all people in your Contacts app.

Your list will look something like **Figure 2.56**.

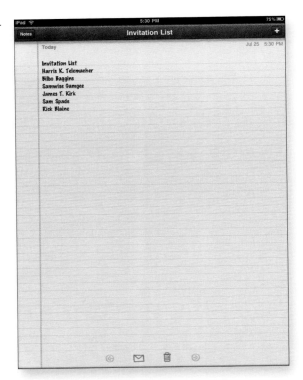

Figure 2.56 The invitation list in Notes.

If you look at the bottom of the invitation list in Figure 2.56, you'll see several icons, including Previous and Next arrows (for paging backward and forward when you have two or more notes), an envelope, and a trash can. The trash can allows you to delete a note. (Be cautious, because after you confirm the deletion by tapping the red Delete button that you see after tapping the trash-can icon, you can't undelete a note.) The envelope icon creates a blank email, with the text of your note in it, ready for you to add a recipient and send, as you do next.

5. Tap the envelope icon at the bottom of your note.

   You see an email similar to the one shown in **Figure 2.57**.

Figure 2.57 The invitation list as an email, ready to send.

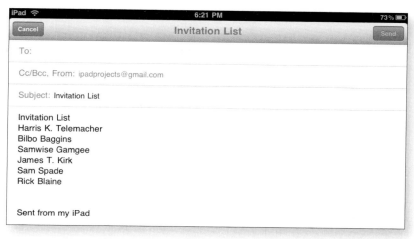

6. In the To field, enter an email address of yours from an account you can access on your iPad; then tap the Send button in the top-right corner to send the email to yourself.

Having the invitation list in an email will make your job easier later in this project, when you're ready to email the invitees your beautifully designed invitation. Next, though, you're going to work on a shopping list.

Make a shopping list

When you have a party, you almost always have some special shopping to do, not only for food and drink, but also for tableware and serving supplies. You may want to start by planning your menu, using a recipe app like Epicurious (which we cover in the **iPad Chef Project** earlier in this chapter), because the app lets you select recipes and then generate a shopping list of the ingredients and supplies you need to purchase.

Alternatively, you can use Notes to create a list of your own and take it with you as you shop. You may want to add street addresses for businesses or a link to a map.

Creating a shopping list in Notes:

1. Tap the Notes icon to open the app.

2. Tap the **+** button to create a new blank note, if necessary.

3. If the digital keyboard isn't visible, tap the top line of the blank note to make it appear.

4. Type **Shopping List** on the first line and then tap the Return key on the digital keyboard.

5. Add a few items for your party shopping list.

   For this task, enter the following shopping list, which you can see in **Figure 2.58**. Be sure to add an extra return after every item to make the list easier to read.

   8 chicken breasts, halved
   6 large peaches
   2 lbs. cherry or grape tomatoes
   Several bunches of fresh herbs
   2 jalapeño chiles
   2 Persian cucumbers
   4 pints fresh raspberries or blackberries
   10 ounces unsweetened baking chocolate
   $1/2$ lb. feta
   25 double-strength paper plates
   25 plastic glasses
   50 napkins
   25 plastic forks, knives, spoons

Figure 2.58
Shopping list.

Chapter 2: Working and Playing in the iPad   103

You want to pick up the produce at the Pike's Place Corner Produce Market, so you're going to add a Produce heading and an address to the list.

6. Tap and hold just before the 6 in *6 large peaches*.

   You see the selection magnifying glass.

7. With the help of the magnifying glass, position the cursor just before the 6.

   When you lift your finger, you see the selection buttons shown in **Figure 2.59**. This time, you can ignore them; you're going to be editing.

Figure 2.59 Shopping list with selection buttons.

8. Tap the Return key twice to create two additional blank lines.

   You see three blank lines after *8 chicken breasts, halved*.

9. Tap the middle blank line to position the cursor there.

10. Type **Corner Produce Market 1500 Pike Place #12 Seattle WA**.

    The address turns into a hypertext link (**Figure 2.60**).

Figure 2.60 The Corner Produce Market as a link in Notes.

11. Tap the link.

    The iPad's Maps app launches and pinpoints the location for you (**Figure 2.61**).

Figure 2.61 Maps showing the location of the linked address.

Now that the shopping list is on your iPad, you can take the iPad with you and delete items as you purchase them, get a map of unfamiliar locations that are already linked for you in Notes, or even use the envelope icon at the bottom of every note to email the list to yourself or to someone who's helping you with your party.

Create your invitations

Another task involved in planning every party is creating and dealing with invitations. Pages for iPad includes an easily modifiable invitation template. In the following tasks, you'll start with that template and then add the information about the party, embed a map showing the party location, and save the template as a PDF file so you can easily email the invitations and track the RSVPs.

Using Pages for iPad to create an invitation:

1. Tap the Pages icon to open the app.

2. If Pages opens to a file instead of the My Documents screen shown in **Figure 2.62**, tap the My Document button in the top-left corner of the Pages screen to go to My Documents.

   Depending on how much you've used Pages, your My Documents screen may look different from the figure.

Figure 2.62
My Documents screen in Pages.

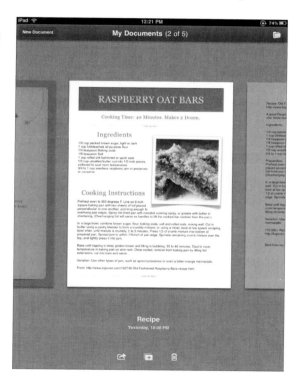

3. Tap the New Document button in the top-left corner.

   You see the Choose a Template screen (**Figure 2.63** on the next page).

Figure 2.63 Choose a Template screen.

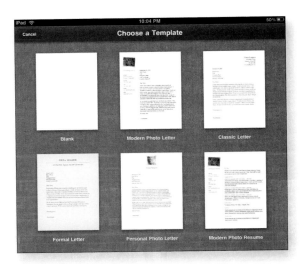

4. Scroll down to the next-to-last row and tap the Party Invite template.

   A document based on the template opens (**Figure 2.64**).

Figure 2.64 Blank Party Invite document.

Chapter 2: Working and Playing in the iPad   **107**

If you look at the new document, which Pages has automatically named Party Invite, you see a paragraph of fake Latin, beginning with *Lorem ipsum dolor,* below the blue *It's Party Time!* heading. You want to change that so that it talks about your party.

5. Tap the *Lorem ipsum dolor* text rapidly three times to select the entire paragraph (**Figure 2.65**).

Figure 2.65 Selected text in the Party Invite document.

6. Type the following replacement text:

    The Sidney family would like to invite you to our housewarming party. Please bring a guest and join us in celebrating our new home.

7. Tap at the very end of the line of text that begins with *Date*.

    A blinking blue cursor appears at the end of the line.

8. Press the Delete key on the digital keyboard to backspace and remove the existing date.

9. Type a new date—for this project, October 7, 2010.

10. Double-tap the text *1:00pm* to select it (**Figure 2.66**).

    A Paste button appears, but you can ignore it. You're going to replace the selected text by entering a new time with the digital keyboard.

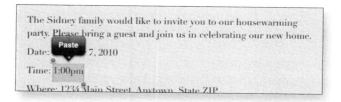

Figure 2.66 Time selected, ready to be replaced by new text.

11. Enter **3:00pm** instead.
12. Replace the address with one of your choosing, if you want.
13. Change the email address in the RSVP to sidneys@example.com.
14. Tap *The Johnsons* three times rapidly to select the entire paragraph and type **The Sidneys** to replace it.

You should end up with something very like **Figure 2.67**.

Figure 2.67 The Party Invite document with most of the text customized.

Next, you'll replace the image in the party invitation with a map. First, though, you need to use the iPad's Maps app to find the party's location.

Chapter 2: Working and Playing in the iPad   **109**

Locating an address with Maps:

1. Tap the Maps icon to open the app.

   The screen you see depends on what you last used Maps for. Although it may not look exactly like the one in **Figure 2.68**, it will have a Search button, a Directions button, and a Search field at the top.

    If you've set your Maps app to display one of the other views, such as Hybrid or Satellite, tap the bottom-right corner and then tap Classic.

Figure 2.68 Maps app's Search screen.

2. Tap the Search button.

   The button turns a darker shade of gray and the map fades a bit to let you know you're in search mode.

3. Tap the Search field in the top-right corner of the Maps screen.

   A blue cursor appears in the Search field, and the digital keyboard opens at the bottom of the screen (**Figure 2.69** on the next page).

Figure 2.69 Maps app, ready to search for a location.

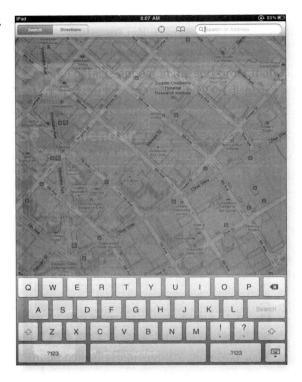

4. Enter an address.

   In this task, you're going to search for the Sidneys' address so you can include a map with the party invitation. They live in an apartment building at 500 Wall St., Seattle, WA 98121, so enter that address.

5. Tap the Search button on the digital keyboard.

   Maps locates the address and pinmarks the location (**Figure 2.70**).

Figure 2.70 A mapped location, marked by a red pin.

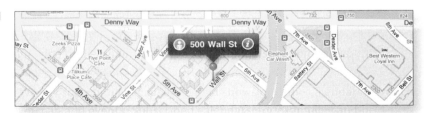

The next thing you need to do is take a screen shot of the map.

6. With the map location still open, press and hold the iPad's Home button, and at the same time, briefly press and release the Power button on the top edge of the iPad's body.

   If you're successful, you see a brief flash of light and hear a sound like a camera shutter. The screen shot is automatically saved to the Saved Photos album of the Photos app.

 The timing can be a little tricky, and this maneuver does take a bit of practice. If the timing isn't quite right, your iPad will go into sleep mode; just wake it up and try again.

Now you just have to place the map in the party invitation you created in "Using Pages for iPad to create an invitation" earlier in this project.

Embedding a map in the invitation:

1. Tap the Pages icon to open the app.

2. If you don't see the My Documents screen, tap the My Document button in the top-left corner.

   If you're completing the tasks in this project in order, you should see the party invitation that you created earlier in the project (refer to Figure 2.67).

3. Tap the camera icon in the bottom-right corner of the placeholder image in the Party Invite document.

   A blue selection rectangle appears around the image, and the Photo Albums list from the iPad's Photos app opens in a floating window (**Figure 2.71** on the next page). (Your album list won't match the one shown in the figure.)

4. Tap Saved Photos in the Photo Albums list.

   You see the images that have been saved to your iPad (**Figure 2.72** on the next page). Your Saved Photos list may include different images but should include the Maps screen shot from the preceding task.

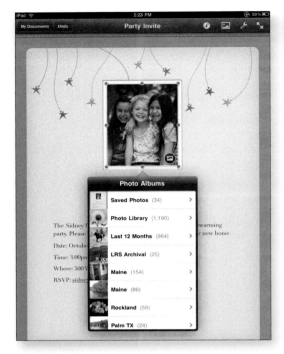

Figure 2.71 Party Invite document with selected image and Photo Albums list.

Figure 2.72 Party Invite document with Saved Photos displayed.

5. Tap the thumbnail of the Maps screen shot.

   After a brief hesitation, the placeholder image of the children is replaced by the map (**Figure 2.73**).

Figure 2.73 Party Invite document with added map.

The Maps screen shot is a much larger image than you really need. Pages lets you mask the image by adding a white border to crop out the parts you don't need. In the following steps, you're going to include just enough of the map to orient party guests.

6. Double-tap the map image.

   A masking slider appears below the map (**Figure 2.74**).

Figure 2.74 Map image with masking slider.

7. Drag the slider to adjust the image's size inside the white mask (**Figure 2.75**).

Figure 2.75 Map image in the Party Invite document showing masking controls.

8. Double-tap the image to center it within the mask, if necessary.

   Try to keep the red pin in the center of the image.

9. Drag the blue selection handles to make the mask larger or smaller, if necessary.

10. When the image is masked as you want it, tap the Done button on the slider.

    The slider disappears, and you should see a final result that's very similar to **Figure 2.76**.

Figure 2.76 Final version of the Party Invite document, with map.

Now all you have to do is save the Pages document as a PDF file so that you can attach it to emails.

Saving the invitation as a PDF file:

1. Tap the Pages icon to launch the app, if it isn't already open.

2. Tap the My Documents button in the top-right corner if you're not seeing the My Documents screen.

3. Locate the Party Invite document in My Documents (**Figure 2.77**).

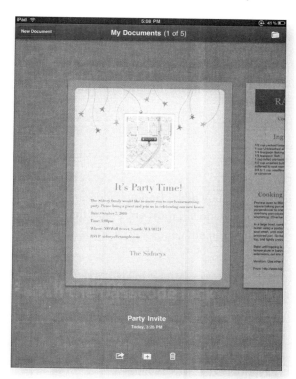

Figure 2.77 The Party Invite document in My Documents, ready to export and mail.

4. Tap the curved-arrow Export icon at the bottom of the screen.

   You see the Export floating window (**Figure 2.78**).

Figure 2.78 Export floating window.

5. Tap Send via Mail.

   The Send via Mail dialog opens, offering three ways to export the invitation (**Figure 2.79**).

Figure 2.79 Send via Mail options.

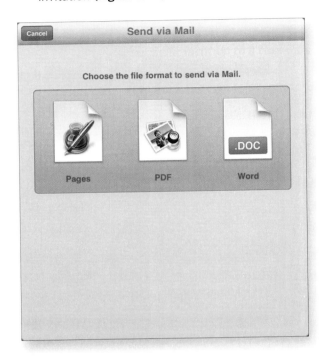

6. Tap PDF.

   You see a brief message that Pages is creating the document; then you see the document embedded in a blank email message (**Figure 2.80**).

Figure 2.80 Party invitation embedded in an email, ready to address and send.

7. Complete and send the message.

   You could add a short note at the top of the body of the email form, address the email to yourself, and add your invitees' email addresses to the Bcc line so that everyone will get a form-letter invitation. This method is efficient but perhaps not as personal as you would prefer.

 You can display the Bcc field in an email by tapping the Cc/Bcc line. Using Bcc means that you protect the privacy, and the personal email addresses, of your recipients.

   If you want to add a short individual note to each invitation, the solution is to email this version to yourself and then create a new email for each person in your invitation list—the list you conveniently emailed to yourself in "Creating an invitation list in Notes" earlier in this project. Having that list of people in your email app makes personalizing and sending invitations a snap.

Handle the responses

If you've created a list of invitees in Notes, you can easily keep track of responses by editing that invitation list as RSVPs arrive by email.

 Because Notes will sync with Microsoft Outlook in Windows and Mail in Mac OS X, you can always have access to the current version of your invitation list.

Tracking your RSVPs:

1. Open your Invitation List note in the Notes app (refer to Figure 2.56).

   You should have entered this list in "Creating an invitation list in Notes" earlier in this project.

2. Delete the names of those who responded by saying that they can't make it to the party.

3. Put an asterisk by the names of those responded by saying they will attend.

4. Add the names of guests or family members who will be coming to the party but weren't in the original list of invitees.

Figure 2.81 shows an example of using an invitation list to track RSVPs.

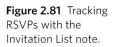

Figure 2.81 Tracking RSVPs with the Invitation List note.

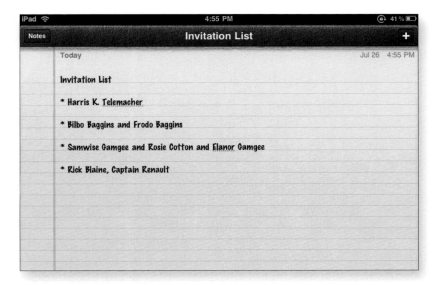

Flash Card Project

Difficulty level: Easy

Software needed: Keynote for iPad ($10), Free Translator 50 (free)

Additional hardware: None

Remember flash cards? Those old-school handheld teaching devices? Well, in this project, you're going to make some.

But put down the posterboard, the scissors, the glue stick, and the colored markers, because you won't be needing them. Instead, you'll make your flash cards on the iPad, using one for-pay app and one free app.

The object of this particular project is to create a set of flash cards to teach words and phrases in another language, but the project really is open-ended. You can use the techniques you develop here to create flash cards on a variety of subjects.

Get your apps in gear

Let's start by getting the apps you need. Their icons are shown in **Figure 2.82**.

Figure 2.82 The two apps you need for this project.

Keynote

Free Translator 50

The central app required for this project is Apple's Keynote. Keynote for the iPad is a scaled-down version of the Keynote application that Apple includes as part of its iWorks productivity suite of applications for the Macintosh. In its Macintosh incarnation, Keynote is a powerful alternative to the widely used Microsoft PowerPoint presentation application. Although the iPad incarnation of the software is somewhat less powerful (and considerably less expensive), you can still make surprisingly sophisticated presentations with it.

In this project, you use Keynote to create a small deck of simple flash cards—a task well within its capabilities.

The second app is Free Translator 50, which (as its name suggests) is free. With it, you can translate single words, phrases, whole chunks of text, and even complete Web pages between any 2 of the 50 languages it supports.

Free Translator is a front end to the translation services provided by Google, Inc. You don't really need the app if you're comfortable using those services manually, but the app is both free and convenient.

Acquiring Keynote and Free Translator 50:

1. Purchase Keynote from the App Store, either directly on your iPad or via iTunes on your computer.

    Just in case you have trouble finding it (although you probably won't), you can open the following URL in a Web browser on your computer, which opens iTunes and takes you right to the app:

    http://itunes.apple.com/us/app/keynote/id361285480?mt=8

2. Download the Free Translator 50 app from the App Store, either on your iPad or via iTunes on your computer.

    You can also enter the following URL in your computer's Web browser to go right to it:

    http://itunes.apple.com/us/app/free-translator-50-more-than/id326079517?mt=8

As soon as you have these apps installed on your iPad, you're ready to begin.

This project requires you to move back and forth between Keynote and the Translator app several times. To make all this navigation easier, we suggest that you put both apps in the same home screen on your iPad. (You move between these apps and Safari as well, but because Safari is in your iPad's Dock by default and visible in every home screen on your iPad, you don't have to worry about moving Safari.)

Translate some words and phrases

The flash-card deck you build can be as thick as you like (or as the storage of your iPad can accommodate). We're going to limit ourselves to three common and useful phrases here, but if you feel ambitious, feel free to add as many as you like. The language we'll use is Italian.

Here are the phrases:

- Hello. My name is _____.
- Where is the restroom?
- Excuse me. Would you take a picture of us?

Translating with Free Translator 50:

1. Tap the Free Translator 50 app's icon, and decline the request to purchase the ad-free version.

   When the request is dismissed, you see the app's interface—and, of course, an ad.

2. At the bottom of the screen, tap Text.

   The text-entry screen appears. At the top are the controls you use to choose the source and destination languages (**Figure 2.83**). The arrow between the two languages shows the direction in which the translation will take place; tap it to change the direction. By default, the source language (on the left) is English.

Figure 2.83 Pick a language, any language.

3. To change the destination language to Italian, tap the current destination language (on the right); scroll through the selection dialog that appears; and tap Italian.

   After you make your selection, the dialog closes.

4. If, for some reason, English *isn't* set as the source language, tap the source language (on the left) and then tap English in the selection dialog.

5. In the large text box of the text-entry screen, type a phrase to translate—in this case, **Hello. My name is _____.**—and then tap the blue Translate button in the top-right corner (refer to Figure 2.83).

   Free Translator displays the text you typed and its translation, as shown in **Figure 2.84**.

Figure 2.84 A phrase and its translation, ready to be copied and used.

6. Tap Text in the bottom-left corner of the screen and repeat step 5, this time typing the second phrase in our list of phrases: **Where is the restroom?**

   Notice that the app keeps the first phrase you translated and adds the second phrase below it. Until you clear the list of translations by tapping the Trash icon, the phrases remain, even when you leave the app and return to it later.

7. Tap Text in the bottom-left corner of the screen and repeat step 5 one more time, this time entering the third phrase in our list: **Excuse me. Would you take a picture of us?**

8. Press the iPad's Home button to close Free Translator.

   You're done with the app—for now.

Collect some illustrations

All you really need for your flash cards in this project are Italian phrases and their translations, but flash cards are far more fun and attractive if you can spice them up with some illustrations. In this section, you use Safari to find and collect appropriate illustrations via Google's image search.

Gathering images with Google Images:

1. In Safari, go to www.google.com/images.

2. On the Google Images page, type **il bagno** in the search field and then tap Search Images.

    Google displays a page of image results (**Figure 2.85**). In case you haven't guessed, *il bagno* is Italian for *bathroom,* and in fact, most of the images shown are of Italian bathrooms. You could have searched for *bathroom,* of course, but why not be authentic?

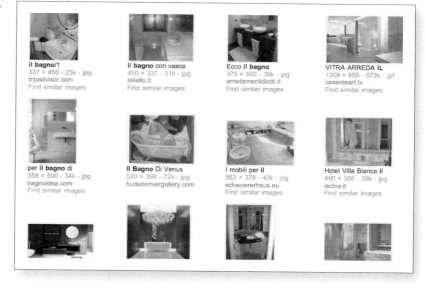

Figure 2.85 Many pictures of Italian bathrooms, courtesy of Google Images.

3. Tap an image that you think would make a good illustration.

    Google displays a split screen, with the Web page on which the image appears at the bottom and a narrow Google search page at the top (**Figure 2.86** on the next page).

Figure 2.86 Google shows the image in context, along with information about the image and some additional options.

4. In the Google search page at the top of the screen, tap the See full size image link.

   The image appears by itself in Safari.

5. Tap and hold the image.

   Two option buttons appear (**Figure 2.87**).

Figure 2.87 You can save any image from the Web in your iPad's Saved Photos.

6. Tap Save Image.

   The image is saved in the Saved Photos album in your iPad's Photos app.

7. In the top-left corner of the Safari screen, tap the Back navigation button.

   The Google split-screen display reappears. Notice that the Google page at the top includes a search field (you can see it in Figure 2.86). You'll use this field to perform additional image searches.

8. Delete your previous keyword in the search field, search for *name tag*, and save a suitable image from the search results.

Chapter 2: Working and Playing in the iPad

9. Search for *photography,* and choose and save a suitable image.
10. Search for, choose, and save an image of the Italian flag.
11. Press the iPad's Home button to close Safari.

Create your flash-card deck

Now that you've gathered all the text and images you need, you're ready to create your flash-card deck with Keynote.

Keep in mind as you work through this section that Keynote automatically saves all your additions and changes, so you can leave and come back to Keynote without worrying about saving your work or losing your place.

 Keynote works only in landscape orientation. If you turn your iPad to portrait orientation, Keynote remains in landscape orientation.

 A terminology note: Keynote creates and displays what it calls *slides*. Although we've been talking about flash cards, we'll use the term *slides* when we discuss flash cards in the context of Keynote.

Creating a new presentation:

1. Launch Keynote, and in the top-left corner of the screen, tap New Presentation.

   The Choose a Theme screen appears (**Figure 2.88**).

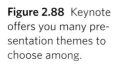

Figure 2.88 Keynote offers you many presentation themes to choose among.

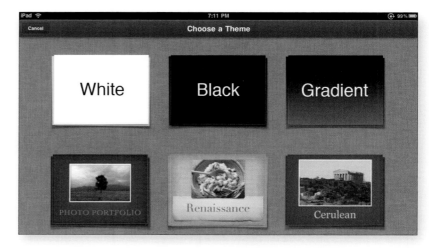

2. Tap the Renaissance theme.

   A new presentation using the theme appears, displaying a placeholder image provided by that theme and two text fields (**Figure 2.89**). Tap a field to cut, copy, or delete it; double-tap a field to edit its contents.

Figure 2.89 The new presentation, with placeholder image and fields.

3. Double-tap the large text field below the picture, and type **Italian Words and Phrases**.

4. Tap the small text field at the bottom of the slide to produce a set of options (**Figure 2.90**).

Figure 2.90 Tap a field to cut, copy, or delete it.

5. You don't need the field at the bottom of the first slide, so tap Delete to get rid of it.

Chapter 2: Working and Playing in the iPad   **127**

6. In the bottom-right corner of the placeholder picture in the center of the slide, tap the camera icon.

   Your iPad's Photo Albums screen opens (**Figure 2.91**).

Figure 2.91 Pick a picture from a photo album to put on the slide.

7. Tap Saved Photos, swipe through the displayed photos until you see the Italian-flag image (which will be on your iPad if you completed the steps in "Gathering images with Google Images" earlier in this project), and then tap it.

   The Italian flag replaces the placeholder picture.

8. Drag the picture until it's centered on the slide above the text (**Figure 2.92**).

Figure 2.92 The replacement picture is in place.

As you drag the picture, guidelines appear when the image approaches the center of the slide vertically or horizontally. Keynote provides these guidelines to help you place objects on a slide more precisely. When the guidelines appear, you can lift your finger; the picture snaps into place.

Now that you have a title slide for the flash-card deck, you can use it to create the other slides.

Duplicating and editing a slide:

1. In the column on the left side of the screen, tap the thumbnail of the slide once to select it, and then tap it again to display the image-editing options (**Figure 2.93**).

   The thumbnails in the left column show the order of slides in your presentation. You can drag them around, edit them, and skip them.

Figure 2.93 Slide-editing options in Keynote.

2. Tap Copy, tap the thumbnail again, and then tap Paste.

   A second thumbnail appears in the column below the first, and its contents are displayed in Keynote's main viewing area. As you'd expect, this image is a duplicate of the first slide—for the moment.

3. Double-tap the text field on the slide, and enter the new text **Hello. My name is _____.**

   When you begin editing a text field, you can use the common iPad editing tools:

   - Tap to select a location in the text.

   - Tap and hold to see a magnified view of the text and then drag to position the cursor.

   - Double-tap a word to select it and then drag the selection handles to adjust the selection.

   - Tap a selection or a location in the text to open the text-editing controls shown in **Figure 2.94**.

Figure 2.94 Text-editing controls for a slide.

4. Tap the picture once, pause, tap again to display the image-editing options, and then tap Delete.

   You want to put a different picture on the slide, so you need to get rid of the existing one first.

5. In the top-right corner of the screen, tap the picture button; then, in the resulting screen (**Figure 2.95**), tap Media.

Figure 2.95 You can add shapes, charts, tables, or your own pictures.

The screen that you see when you tap the picture icon gives you access to a lot of graphic objects that you can place in slides. The Media collection shows your photo albums.

6. Tap Saved Photos and then tap the name-tag image (which will be on your iPad if you completed the steps in "Gathering images with Google Images" earlier in this project).

7. Drag the photo so that it's centered above the text field.

You have the English version of a phrase slide. Next, you make the slide that has the Italian translation.

Making an Italian slide and completing the deck:

1. Follow steps 1–2 of the preceding task, "Duplicating and editing a slide," to make a duplicate of the English slide you just completed.

2. Press the iPad's Home button to return to the home screen.

3. Launch the Free Translator 50 app.

    You see the phrases that you translated earlier. You want to copy the Italian translation of the English text that currently appears on the duplicate slide.

4. Below the Italian text that you want, tap the Copy button.

    The Copy button copies the Italian text to your iPad's clipboard.

5. Press the iPad's Home button to return to the home screen.

6. Launch Keynote.

    The slide that you just duplicated is displayed.

7. Double-tap the text field to edit the text, select all the text, and paste the Italian translation that's on the clipboard.

    The English text on the duplicated slide is replaced with the Italian translation. Now you have an Italian slide to match the English one. Next, you'll duplicate *this* slide and make it into the next English slide in the presentation.

8. Repeat all the steps in "Duplicating and editing a slide," but with the following changes:

    - In step 3, enter the second English phrase: **Where is the bathroom?**
    - In step 6, select the bathroom image.

9. Make an Italian slide by repeating steps 1–7.

10. Repeat steps 8–9, substituting the appropriate pictures and text (in step 3, enter **Excuse me. Would you take a picture of us?**), to make the third set of slides.

Voila! You have a short flash-card presentation of Italian words and phrases. If you like, you can add to the presentation at any time.

Viewing the presentation:

1. In the thumbnail column on the left side of the Keynote screen, tap the topmost thumbnail.

   The topmost thumbnail represents the title slide that you made in "Creating a new presentation" earlier in this project. Keynote displays the title slide in the main work area.

2. Tap the Play button in the top-right corner.

   The slide expands to fill the entire screen.

3. Tap anywhere on the slide.

   The next slide is displayed.

4. Continue to tap until the last slide is displayed.

5. Tap the last slide.

   Tapping the last slide ends the presentation and displays the slide-composition screen again.

You can double-tap a presentation at any time to return to the slide-composition screen.

Now that your flash-card composition is complete, it's ready to share. Naturally, you can share it simply by using your iPad to display it, but if you want to give it to someone, you need to export it.

Exporting the flash cards:

1. In the top-left corner of the Keynote screen, tap My Presentations.

   Keynote displays the available presentations, with a large thumbnail of the flash-card presentation centered onscreen. You need to rename the presentation, because it currently has a generic name (such as Presentation 1).

2. Tap the presentation's name, delete the existing name in the editable title field that appears, and type a new name.

3. Tap the thumbnail to complete the renaming.

4. Tap the Export button.

   A dialog offers you three options:

   - You can send the presentation via email. If you choose this option, you can send either a Keynote file or a PDF version of that file.

   - You can share the presentation on the iWork.com site so that other people can download a copy, and you can share it as either a Keynote file or a PDF. When you choose this option, Keynote creates an email with the download information that you can send to your intended recipients.

   - You can export the presentation in either Keynote or PDF form to your iPad's file-sharing area and then use iTunes to copy the file from your iPad to your computer. This process is explained in the **File Management Project** earlier in this chapter.

5. Tap the option of your choice, and follow the onscreen instructions.

Vacation Planning Project

Difficulty level: Easy

Software needed: KAYAK (free), TravelTracker (free)

Additional hardware: None

There's nothing like getting away from it all. Unfortunately, most vacations are nothing like getting away from it all. Usually, you end up taking a lot of it all with you—your clothes, toiletries, reading material, cameras, chargers, cables, and so on. And when you *do* leave, some of what you're getting away from still haunts you, as you find yourself waiting to board your flight thinking, "Did I remember to take out the trash and start the dishwasher?"

What's more, to get away from it all, you first have to dance the airline online tango, stepping from one airline Web site to another, searching for a flight that you can afford but that doesn't involve three intermediate stops and a 3 a.m. departure time.

In this project, you use two free apps that help ameliorate the stress of getting away to be unstressed. With them, you can find the right flights and manage the trivia of planning your voyage away from it all—or almost all, because you're going to take your iPad with you, right?

Don't forget the charger.

Pack your apps

The first app you need is TravelTracker. This app has existed in one form or another since back in the 20th century, when Apple made the proto-iPad that it called the Newton. Through many incarnations, it has helped travelers track their traveling trivia: the itineraries, the frequent-flyer miles accrued, the places to visit, the stuff to pack, the things to do, and the places to stay. No, it won't magically handle all these things for you—you still have to enter these bits of information into the app—but it keeps them in one place, easily available and nicely arranged.

The second app is KAYAK (technically called KAYAK Explore + Flight Search in the iTunes Store). This app, and the associated KAYAK Web site for which the app is a convenient front end, links with most major airlines. With it, you can search for flight information, and pick the flights that fit your budget and your traveling preferences.

The icons for the apps are shown in **Figure 2.96**. Go and get them.

Figure 2.96 The two apps you need for this project.

TravelTracker   KAYAK

Getting the apps:

1. Download the free TravelTracker app from the App Store, either directly on your iPad or via iTunes on your computer.

   You can enter the following URL in your computer's Web browser to have it open iTunes and take you right to the download page:

   http://itunes.apple.com/us/app/traveltracker-personal-travel/id284918921?mt=8

2. Download the free KAYAK app from the App Store, either directly on your iPad or via iTunes on your computer.

   Enter the following URL in your computer's Web browser to have it open iTunes to the app's page in the App Store:

   http://itunes.apple.com/us/app/kayak-explore-flight-search/id363205965?mt=8

Set up a trip with TravelTracker

To set up a trip in TravelTracker, you need (of course) a trip to set up. For the purposes of this project, we're going to make up a trip just so we have something to illustrate when we walk you through the steps. Feel free to use your own trip details, if you prefer, substituting your entries for what we show here.

Our imaginary trip is from Los Angeles to Portland, Oregon, for the holiday season. The visit starts on December 21, 2010, and ends on December 28, 2010. (If you're reading this book after those dates, you'll need to adjust the dates accordingly.) Near the end of the trip, we'll plan a post-Christmas dinner at an exclusive restaurant to thank our hosts for putting us up.

In this part of the project, you create a new trip and schedule the dinner.

Creating a new trip:

1. Launch the TravelTracker app on your iPad.

   The first time you open the app, it provides a helpful tip, as shown in **Figure 2.97**. The app tells you about the screen you're looking at and mentions another service, TripIt, that makes TravelTracker even more useful. We won't use TripIt in this project, but it's worth checking out.

Figure 2.97
The All Trips page in TravelTracker, complete with a tip.

2. If the tip appears, tap OK to dismiss it; then tap the New Trip button in the top-right corner of the screen.

   A New Trip form appears (**Figure 2.98** on the next page). Note that the form starts the trip on the current date. You need to change that date, but first, you should give the trip a name.

Figure 2.98 Creating a new trip.

3. In the form, tap Title.

    An Edit Trip Name form appears.

4. Type a title, such as **Holiday In Portland** (**Figure 2.99**).

Figure 2.99 Every trip needs a name.

5. Tap the Save button in the top-right corner to save your work and close the form.

    The New Trip form returns.

6. Tap Start Date.

    An Edit Trip Dates form opens, featuring a standard iPad date-selection widget.

Chapter 2: Working and Playing in the iPad   137

7. Tap the Start field, and use the widget to dial the start date—for this project, December 21, 2010.

   Your selected date appears in the Start field (**Figure 2.100**).

Figure 2.100 Select your trip dates with the date-dialing widget.

8. Tap the End field, and use the widget to dial the end date—for this project, December 28, 2010.

   The selected end date appears in the End field.

9. Tap Save in the top-right corner to save your work and close the form.

   The New Trip Form returns.

10. Tap Save.

    The trip is saved, and you see a blank itinerary screen for the Holiday In Portland trip (or whatever you called the trip you just created), as shown in **Figure 2.101**.

Figure 2.101 The Holiday In Portland trip has a blank itinerary for now.

If you're using TravelTracker for the first time, a help tip explains the itinerary screen that you just created. This itinerary screen holds various items for your trip, such as flights, planned meals, and lodging—or *will* hold these items when you create them. Right now, it's stunningly empty.

It's time to put something in your itinerary. Something tasty.

You can buy add-ons that expand TravelTracker's capabilities. One such add-on, which currently costs 99 cents, provides checklists for items to pack and pretrip tasks to perform (such as stopping mail delivery). If you really do find yourself worrying about whether you remembered to start the dishwasher as you stand in line waiting to board, that add-on may be just what you need.

Scheduling a dinner:

1. In the top-right corner of the itinerary screen (refer to Figure 2.101), tap New Item.

   The New Item screen displays a list of items that you can add to the itinerary (**Figure 2.102**).

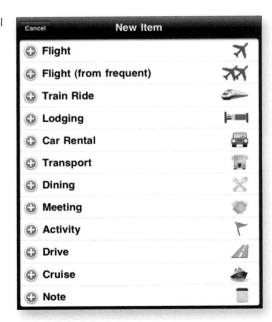

Figure 2.102 Any or all of these items can go in your itinerary.

2. Tap Dining.

   A New Dining screen appears (**Figure 2.103**). This screen, where you add restaurant and reservation information, sets the reservation to the first day of the trip by default. The time is also set by default to 6 p.m.; you won't need to change it for this task. (6 p.m. is a good time for dinner.)

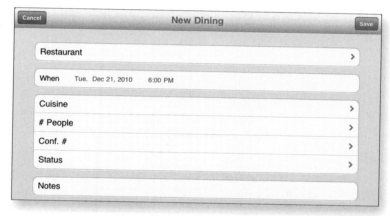

Figure 2.103 Enter information about your dining plans here.

3. Tap the When field.

   A date and time selection widget appears, displaying the default date and time (**Figure 2.104**).

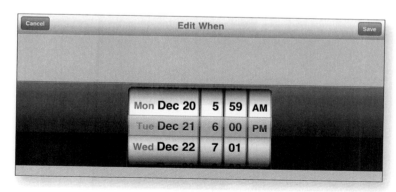

Figure 2.104 Set this widget so you won't miss dinner.

4. Set the date to December 27 and then tap Save.

   You return to the New Dining screen.

5. Tap the Restaurant field.

   A blank Edit Name screen appears (**Figure 2.105** on the next page).

Figure 2.105
In TravelTracker, the restaurant name seems to include phone, address, and city as well.

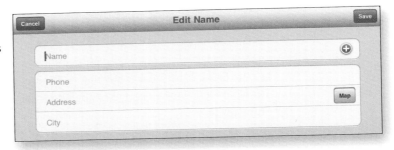

6. Tap each field, and enter some appropriate information (using a real restaurant or just making something up).

 If you happen to have the restaurant in your iPad's Contacts app, you can tap the blue plus (+) icon at the right end of the Name field to add the restaurant's name, address, phone number, and city to the Edit Name screen. If you don't, you can enter each item manually.

7. Tap Save to save your work and close the Edit Name screen.

The New Dining screen reappears, with the restaurant information filled in.

8. (Optional) Tap each field in the New Dining screen, and enter information for each one.

A typical New Dining screen looks like **Figure 2.106** when it's filled out.

Figure 2.106 Dinner at an exclusive restaurant, ready to be added to your itinerary.

Chapter 2: Working and Playing in the iPad   **141**

9. Tap Save.

   Your itinerary now has a dining item scheduled (**Figure 2.107**).

Figure 2.107 At last, your itinerary has something in it.

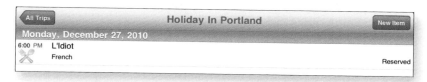

10. Press your iPad's Home button to exit TravelTracker.

    It's time to move to KAYAK and schedule a flight.

Find flights with KAYAK

As you may have noticed in TravelTracker, you can add flights to your itinerary. Before you can do that, though, you have to have some flights to add. You use KAYAK to find those flights and even book them.

 The flight information shown in this section is subject to change and is shown only for example purposes. No endorsement of particular airlines is intended.

Setting flight routes with KAYAK:

1. Launch the KAYAK app on your iPad.

   KAYAK opens, showing you its Explore screen. This screen displays locations around the world, and you can use it to find current deals on travel to various locations. In the next step, though, you're going to skip directly to the Flights screen.

2. Tap the Change Location button in the top-left corner.

   A search form drops down (**Figure 2.108** on the next page).

 You can also set your starting and destination locations in KAYAK's Flight Search screen.

Figure 2.108 Use this form to search for the airport from which you want to depart.

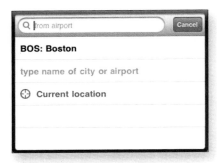

3. In the search field, begin typing **Los Angeles**.

   As you type, airport abbreviations begin to appear below the search field. As soon as you see *LAX*, you can stop typing (**Figure 2.109**).

Figure 2.109 KAYAK knows about a lot of airports.

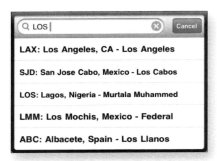

> **tip** Alternatively, you can tap Current Location in the Change Location search form to have KAYAK list the airports in your general area.

4. Tap LAX in the list of airports that appears and then tap Flights in the top-right corner of the Explore screen.

   The Flight Search pane appears. Notice in **Figure 2.110** that KAYAK has made LAX the destination. It's easy to swap the From and To locations, however, as you're about to see.

Figure 2.110
The KAYAK Flight Search pane.

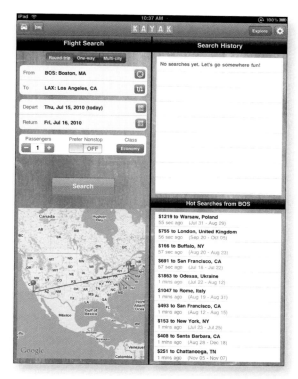

5. At the top of the Flight Search pane, tap Round Trip.

6. If LAX is the To location, tap the swap button; then tap the To location, and in the Choose Destination Airport screen that appears, begin to type **Portland, Oregon** (**Figure 2.111**).

Figure 2.111 Pick a destination airport.

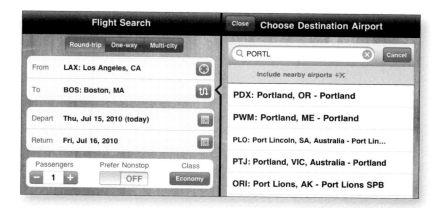

7. When PDX (the airport in Portland, Oregon), appears in the form, tap that entry; then tap Close at the top of the screen.

 Back in the Flight Search pane, PDX is listed as the destination.

Next, you'll set the travel dates and search for a flight.

Setting travel dates and finding a flight:
 1. In the Flight Search pane (refer to Figure 2.110), tap the Depart field to open the Choose Depart Date pane on the right side of the Explore screen.

 2. Swipe down the Choose Depart Date pane to December, and tap 21 on the calendar (**Figure 2.112**).

Figure 2.112 KAYAK uses calendars instead of the date-dialing widget in TravelTracker.

3. In the Flight Search pane, tap the Return field to open the Choose Return Date pane on the right side of the Explore screen.

4. Swipe down the Choose Return Date pane and tap December 28; then close the pane.

5. Back in the Flight Search pane, find the Passengers indicator; tap the + button to increase the number of passengers to 2; and then tap the Prefer Nonstop switch to turn that setting on (**Figure 2.113**).

 You can tap the button below Class to choose other classes of flight. You can choose Business or First Class if you want to pamper yourself and pay considerably more.

Figure 2.113 Set the number of passengers, and tell KAYAK whether you want to go nonstop and what class you want to travel in.

6. Tap the big orange Search button above the map in the Explore screen.

   KAYAK displays the flight search results.

7. Swipe up and down the search results to find a suitable flight, and when you do, tap it.

   The Flight Details pane appears on the right side of the screen, with Booking below it (**Figure 2.114** on the next page).

   At this point, you can call one of the listed numbers or click a link to book your flight. Because this task is just an example, however, make a note of the airlines and the flight numbers (sadly, KAYAK doesn't let you copy the text in the Flight Details pane to the clipboard) so that you can use them in the next section.

8. Press the iPad's Home button to exit the KAYAK app.

Figure 2.114 Click a link or make a phone call to book the flight that you've chosen.

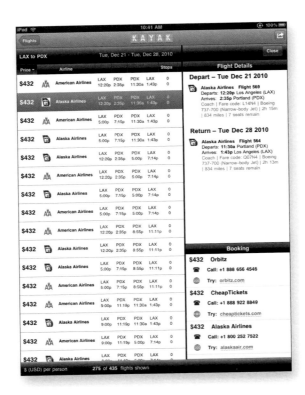

Add flight information to your itinerary

Now that you have your flight information, you can add your departure and return flights to your TravelTracker itinerary.

Entering flight information with TravelTracker:

1. Launch TravelTracker on your iPad again.

    You see your itinerary screen just as you left it (refer to Figure 2.107).

2. Tap New Item in the top-right corner to open the New Item form.

Chapter 2: Working and Playing in the iPad   **147**

3. Tap Flight.

   The New Flight screen appears (**Figure 2.115**).

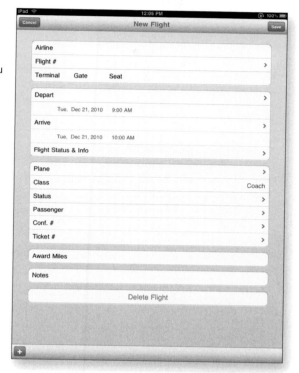

Figure 2.115 The New Flight screen holds a lot of information about your flight, and TravelTracker helps you fill it out.

4. Tap the Airline field at the top of the screen.

   The Select Airline screen appears (**Figure 2.116** on the next page).

Figure 2.116 Select an airline.

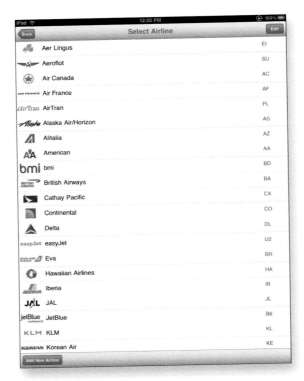

5. Find the airline for your departure flight, and tap it.

 In the unlikely event that your airline isn't listed in the Select Airline screen, tap Add New Airline and follow the onscreen instructions to add it.

The New Flight screen reappears.

6. Tap the Flight # field to open the Edit Flight # screen.
7. Type the flight number (**Figure 2.117**), and tap Save.

Figure 2.117 The flight-number field can hold a *very* long flight number.

The New Flight screen returns briefly but is quickly replaced by the Departure Date screen, which contains a date selection widget.

8. If the departure date for this project—December 21, 2010—isn't already selected, use the widget to dial it in (**Figure 2.118**) and then tap Save.

Figure 2.118 Use the widget to set your flight date.

TravelTracker looks up the flight and fills out the rest of the form, as shown in **Figure 2.119**.

Figure 2.119 TravelTracker fills out the rest of the New Flight form as soon as it has an airline, a flight number, and a date.

9. Tap Save.

    The departure flight appears in your itinerary screen.

10. Repeat steps 2–9, using the return flight number, the airline, and the return date that you got from your KAYAK search (assuming that you completed both tasks in "Find flights with KAYAK" earlier in this project).

    Your itinerary now contains your departure flight, your return flight, and your dinner information (**Figure 2.120**).

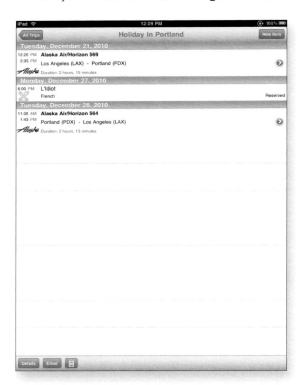

Figure 2.120 Your itinerary is starting to look useful.

11. Press the iPad's Home button to exit TravelTracker.

    The beginnings of your vacation plan are stored in TravelTracker, with an assist from KAYAK. Have a great trip! Don't forget to drop us a postcard!

3

Music, Books, and Movies on the iPad

In the preceding chapter, we take issue with the conventional notion that the iPad is just a media-consumption device. We think we prove our point.

Nonetheless, the iPad really is a delightful media-consumption device if you want to use it as one. The trick is getting the media you want on it, when you want it, and in the form that you want it.

This chapter shows you those tricks. We show you how to sync music and video between your computer and the iPad. We show you alternative ways to get video on your iPad. We show you where to obtain e-books and which apps you can use to read them. We even show you how to make and modify e-books for your iPad.

Our philosophy is this: If you're a consumer, you owe it to yourself to be the best consumer that you can be.

Music Syncing Project

Difficulty level: Easy

Software needed: iTunes

Additional hardware: Mac or PC

When Apple announced its iPad at the beginning of 2010, one of the most common criticisms that pundits aimed at the device was "It's just nothing but a giant iPod touch!"

As if that were a *bad* thing.

In fact, though, this facile critique has some basis: Setting aside the iPad's unique capabilities, it does have much in common with the iPod touch. In particular, your iPad has a built-in iPod that you can use to play the many thousands of songs your iPad can hold.

To act in its secret identity as a giant iPod touch, however, your iPad needs songs to play. If you're among the ever-shrinking number of souls who've never had an iPod—or even if you have one but find the whole music-syncing thing to be mystifying—this project shows you how to get the songs that you want, by the artists that you want, in the styles that you want, from your iTunes Library onto your plus-size iPod touch.

Sync everything

If you have a relatively small iTunes Library (that is, one that can fit easily into your particular iPad's storage space), deciding what music to put on your iPad is simple: Just put your whole Music library on the device, and don't worry about it. Even the smallest-capacity iPad has ample room to hold several thousand songs.

 Apple provides a ballpark estimate of about 250 songs per gigabyte. Going by that estimate, a 16 GB iPad can store 4,000 songs and still have a third of its storage space left for documents, pictures, apps, and other stuff.

Assuming that your Music library fits on your iPad, getting it there is just a few clicks and a sync away.

Syncing your Music library to your iPad:

1. Connect your iPad to your computer.

2. Launch iTunes.

3. In the Source list on the left side of the iTunes window, select your iPad.

4. Click the Music tab at the top of the main pane of the iTunes window.

   The contents of the Music tab for your iPad appear, with the main syncing options laid out at the top (**Figure 3.1**).

Figure 3.1 The main music syncing options for your iPad.

5. Select the Sync Music option, and then, below it, select Entire music library.

6. (Optional) Select Include music videos.

   Keep in mind that each music video takes up considerably more space than a typical song. But if you have the space (see the nearby sidebar), why not?

7. Click the Apply button in the bottom-right corner of the iTunes window.

   iTunes begins copying your entire Music library to your iPad. Depending on the size of the library and the speed of your computer, this process can take several minutes, so be patient; you have to do this only once. When you add more songs to your iTunes Library, only the additional songs will be copied to your iPad the next time you sync.

Making Smaller Songs to Save Space

Songs can be stored in various formats, some of which take up more space than others do. To conserve iPad storage space, do the following

1. With your iPad selected in iTunes' Source list, click the Summary tab.
2. Near the bottom of the tab, select the option titled Convert higher bit rate songs to 128 kbps AAC (**Figure 3.2**).

This option increases the time that it takes to sync your music, because iTunes must convert every song that isn't in 128 Kbps AAC format as it syncs. Also, the process slightly reduces the sound quality of the synced songs on your iPad. But unless you have very-high-quality speakers or headphones attached to your iPad (and *very* good ears), you won't be able to tell the difference in the converted songs' sound quality.

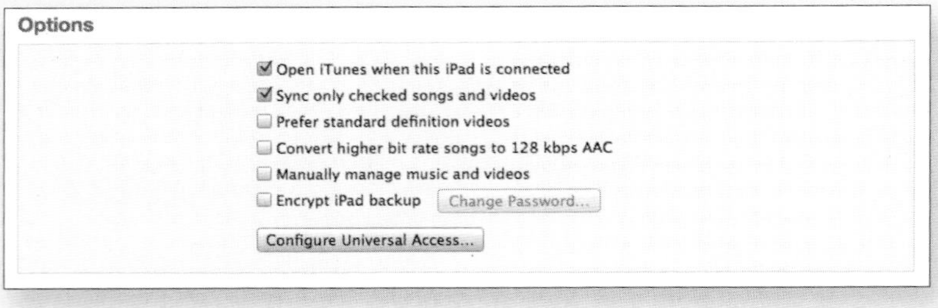

Figure 3.2
The options at the bottom of the Summary tab.

Sync artists and genres

If your iTunes Music library is too large for your iPad, or if you just don't want to copy the whole thing, you can narrow down which songs get synced in several ways. One of the easiest ways is to choose your favorite musical artists and musical genres, and then sync only the songs that match your choices.

iTunes presents the artists and genres associated with your songs in the music syncing tab for your iPad, right below the main music-syncing options (**Figure 3.3**). You use the check boxes in these lists to select your favorites.

Figure 3.3 The Music syncing tab lists your Music library's artists and genres so that you can choose which ones to sync.

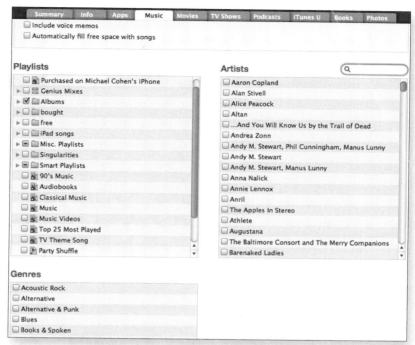

Syncing selected artists and genres to your iPad:

1. Follow steps 1–3 of "Syncing your Music library to your iPad" earlier in this project.

2. Select the Sync Music option, and then, below it, choose the radio button titled Selected playlists, artists, and genres.

   The Playlists list, the Artists List, and the Genres list become accessible so that you can check or uncheck items in them.

   We'll ignore the Playlists list for now, because we cover it in the next section of this project.

> **tip** If some of the items in the Playlist list happen to be checked, you can easily uncheck all of them so that they don't interfere with this part of the project. To do so, hold down the Command (Mac) or Ctrl (Windows) key, and click one selected check-box item to deselect everything in the list. Similarly, clicking an unselected check box with the key held down selects every item in the list. This technique, by the way, works with almost every list in iTunes that has check boxes.

3. In the Artists list, select the artists whose songs you want to sync with your iPad.

If you have a lot of artists listed, you can type part of an artist's name in the search field above the Artists list so that it shows only artists that match what you typed.

4. In the Genres list, select the genres that you prefer.

iTunes uses the genre assigned to the song from the iTunes Store (if it came from there) or from the music vendor from which you obtained the song. It also uses information from an online database when you rip a CD to add its songs to your Music library. You can modify this information if you like; see the iTunes Help topic "Editing Song and CD Information" to learn how to do this.

5. Click Apply in the bottom-right corner of the iTunes window.

   iTunes syncs the songs that match your artist and genre selections to your iPad, and removes any songs from your iPad that don't match your selections.

Make and sync playlists for your iPad

Some people who spend a lot of time with iTunes delight in arranging their songs in all sorts of ways by using iTunes' playlist features. Many other people, however, find playlists abstract and confusing, and shy away from them as though they were a nest of snakes.

If you're comfortable with playlists, simply skip to "Selecting and syncing playlists" later in this project. The rest of you, read on.

Playlists are actually quite simple: They're lists of one or more songs. That's it. They don't contain any actual songs—just references to songs. That's the part that seems to confuse people, because in iTunes, the act of adding songs to a playlist looks very much like copying the actual songs. It isn't. When you add songs to a playlist, you're simply adding *references* to those songs, as **Figure 3.4** illustrates.

Figure 3.4 It may *look* like we're copying four songs into a playlist, but we're really copying references to those songs.

Think of a room full of people. You can make a list of everyone in the room, and you can make another list of, say, just the left-handed people in the room. The people themselves aren't cloned and embedded magically in your lists; the lists contain only references to the people. The people themselves stay in the room where they were all along.

Similarly, when you add a song from your iTunes Music library to a playlist, the song itself isn't copied to the playlist; it stays where it is, in your Music library. You can add the same song to as many playlists as you like, just as you can add the same person to as many lists as you like, but no matter how many times a person is added to a list or a song is added to a playlist, only one physical person or only one actual song is involved.

When you sync a playlist from iTunes to your iPad, iTunes checks the songs that are already on your iPad. If a song in the playlist is already there, iTunes doesn't copy it to your iPad; if it isn't, iTunes copies the song to your device. That way, you can sync as many playlists to your iPad as you want, and even if the same song appears in all those playlists, only one copy of that song goes from your Music library to your iPad.

Playlists appear in the Source list on the left side of the iTunes window. Over time, you may find that you've made dozens of playlists. You can shorten the Source list by creating folders where you can store various playlists.

When you sync playlists between iTunes and your iPad, you can choose individual playlists, folders full of playlists, or individual playlists inside folders.

When you check a folder of playlists in iTunes and sync that playlist folder to your iPad, all the playlists inside that folder are synced to your iPad. The original version of the iPad software doesn't display the playlists on your iPad in folders; instead, the playlists on the iPad appear in a single long list. The iOS 4 version of the iPad software, however, does allow playlist folders on your iPad, and it keeps the playlists in their folders just as they are in iTunes.

In the following tasks, you create a folder for the playlists that you want to have on your iPad, create a playlist inside that folder, add songs to that playlist, and then sync everything. When you have these steps down, you can go through your iTunes Library at leisure and create playlists and folders for your iPad to suit your needs and desires.

Making a playlist and playlist folder in iTunes:

1. In the iTunes Source list, click Music.

   All your music in iTunes appears in a list (**Figure 3.5**).

Figure 3.5 The Music library lives near the top of the Source list.

Before you can create a playlist or a playlist folder, iTunes must be displaying content from one of your libraries.

2. Choose File > New Playlist Folder.

   A folder appears in the Playlists section of the Source list with its title selected, ready for you to edit it (**Figure 3.6**).

Figure 3.6 A new playlist folder awaiting its new name.

3. Type a folder name, and press Return (Mac) or Enter (Windows).

   For this task, type **iPad songs**. When you press Return or Enter, the folder is renamed. If your Playlist section contains other folders, they appear in alphabetical order, and the newly named folder moves to its correct alphabetical position among them.

4. Click the new folder to select it and then choose File > New Playlist.

   An untitled playlist appears inside the folder, ready for you to rename it.

5. Type a new name for the playlist (whatever name you like), and press Return or Enter.

Now that you've successfully created a playlist inside a folder, you can add songs to it. You can add songs to a playlist whether that playlist is in a folder or not, of course, but putting playlists inside folders makes it more convenient to sync them with your iPad. In this project, you're using only one playlist, but you can use the folder you just created to contain all the playlists destined for your iPad.

 You can drag any playlist in the Source list onto a folder to put it in the folder, and you can take a playlist out of a folder by dragging it to the left edge of the iTunes window.

Adding songs to a playlist:

1. Click Music in the iTunes Source list.

   All the content in your Music library is displayed in the main pane of the iTunes window.

2. Click a song to select it.

   You can select multiple songs by Shift-clicking. If your iTunes Library is displaying songs by album cover in iTunes' grid layout, you can select the album by clicking its cover. (To view your Music library in grid layout, choose View > as Grid.)

3. Drag the song to the playlist to which you want to add it.

   The song is added to the playlist (refer to Figure 3.4 earlier in this project).

 tip You can delete a song from a playlist at any time. First, click the playlist to see its contents; then select the song and press the Delete key. Don't worry—the song is still in your Music library.

Now that you have a playlist folder and a playlist to play with, you can sync them with your iPad.

Selecting and syncing playlists:

1. Follow steps 1–3 of "Syncing your Music library to your iPad" in the first section of this project.

2. If they aren't already selected, select the Sync Music option and, below it, Selected playlists, artists, and genres.

3. In the Playlists list, select a playlist folder.

   The Playlists list displays folders before individual playlists, so chances are that your iPad songs folder for this project is near the top, as shown in **Figure 3.7**.

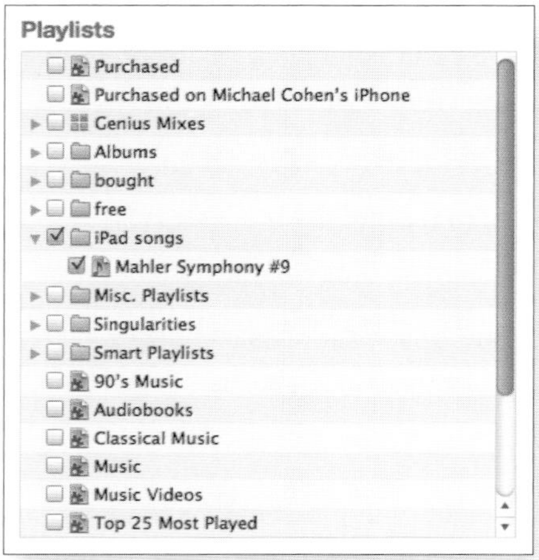

Figure 3.7 Adding a playlist folder for syncing.

4. Click Apply in the bottom-right corner of the iTunes window.

    iTunes syncs your new playlist, along with any other items selected in the three lists in the music syncing tab (refer to Figure 3.3 earlier in this project).

Get Smart

Aside from playlists and playlist folders, iTunes provides smart playlists. When you make a smart playlist (by choosing File > Smart Playlist), you specify the conditions that a song must satisfy to be in that playlist. You could specify all songs with the word *Love* in their titles that have been added to your iTunes Music library in the past year, for example. Whatever songs match those conditions end up listed in the smart playlist. If you add a new song to your Music library that satisfies the smart playlist's conditions, it ends up listed in the smart playlist too.

You can sync smart playlists with your iPad, which is why we're bringing the whole topic up. Consult iTunes Help to learn more about smart playlists.

Create a playlist on your iPad

While we're on the subject of playlists, we should point out that you can make a playlist on your iPad, using any of the songs, artists, albums, and genres there. Any playlist that you make on your iPad syncs back to iTunes, where you can modify it and then sync it back to your iPad.

Making a playlist:

1. Tap the iPod app's icon on your iPad to open it.
2. Tap Music at the top of the Library column.

    Your Music library appears in a list (**Figure 3.8** on the next page).

Figure 3.8 The iPod on the iPad.

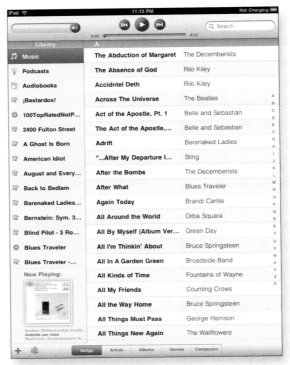

3. At the bottom of the screen, tap Songs; then, in the bottom-left corner, tap the plus (+) icon.

   A New Playlist dialog appears.

4. Enter a name for your playlist, and tap Save.

   We entered the name **From iPad**, but you can use any name you like. When you tap Save, your iPad displays the songs that it contains in alphabetical order and instructs you to add songs to the playlist.

5. Swipe through the list of songs, tapping the ones you want to add to your new playlist.

   As you tap a song's title, it turns gray, indicating that it's been added.

> **tip** You can use the search box at the top of the screen to find the songs you want. You can also use the buttons at the bottom to add albums, artists, genres, and composers.

Chapter 3: Music, Books, and Movies on the iPad   163

6. Tap the blue Done button.

   Your new playlist appears (**Figure 3.9**). Now you can tap any song to delete it from the playlist (such as the duplicate song shown in the figure) or tap the playlist in the Library column to delete the whole thing.

7. Tap Done again.

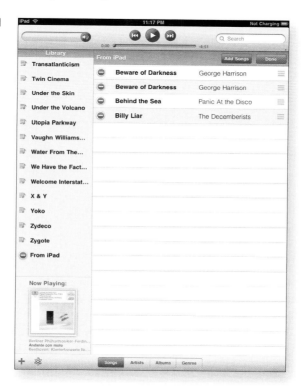

Figure 3.9 A new iPad playlist awaiting final approval.

Manage your music by hand

Some people like total control of every single item on their iPads, preferring to add and remove songs manually rather than rely on syncing. If you're one of those people, this last section and iTunes' Manually manage music and video option are for you.

The aforementioned setting allows you to drag songs, albums, playlists, videos, and TV shows from your iTunes libraries directly to your iPad's Source list, but it stops iTunes from syncing any of these items automatically. If, for example, you have iTunes set to automatically sync movies you haven't finished watching (not covered in this project, but see the **Movie and TV-Show Syncing Project** later in this chapter), that syncing capability is disabled when you choose the Manually manage option. Manually really does mean manually.

Managing music manually:

1. Connect your iPad to your computer.

2. Launch iTunes.

3. Select the iPad in the Source list.

4. Click the Summary tab in the main iTunes window.

5. In the Options section at the bottom of the tab, select Manually manage music and videos (refer to Figure 3.2 earlier in this project).

6. Click Apply in the bottom-right corner of the iTunes window.

Create and Convert E-Books Project

Difficulty level: Intermediate

Software needed: calibre (free)

Additional hardware: Mac or Windows PC

When Steve Jobs introduced the iPad, the feature that created the most buzz was iBooks. Shortly thereafter, Apple announced that iBooks would be free, but iPad purchasers would have to download and install it from the App Store. Some people took this as a cynical attempt to use iBooks as a loss leader and get people into the App Store; the more forgiving considered it to be a way for Apple to spend more time honing iBooks' performance and functionality. Either interpretation is surely plausible (and both may be).

Initially, iBooks supported only an e-book format known as ePub, but the first update to the iBooks app added support for PDF documents. With the wealth of PDF content (user manuals, government documents, books, theses, and so on) available on the Internet and on software CDs and DVDs, your iBooks library can be a repository for significant quantities of reference materials in addition to the books you purchase (or get for free) from the iBookstore.

Make your own PDFs

PDF support greatly simplifies the process of putting your own content on the iPad. Using a third-party PDF-creation tool or Mac OS X's built-in printer support for writing PDF files, converting your documents to PDFs is a simple matter of opening them and choosing the right printer or print option. Drag the resulting PDF into iTunes and sync your iPad to make the PDF accessible on the go.

The following task demonstrates just how simple it is for a Mac user to create a PDF, using the Microsoft Word document for this project as the source.

Creating a PDF:

1. On your computer, open the document in your application of choice.

   For this example, we're using this project file in Microsoft Word 2008.

2. Choose File > Print.

   The Print dialog opens. Ours appears in **Figure 3.10**. Although the basics will be the same, the application and printer you're using could make the dialog sport more or fewer features.

Figure 3.10 Use the Print dialog to create PDFs.

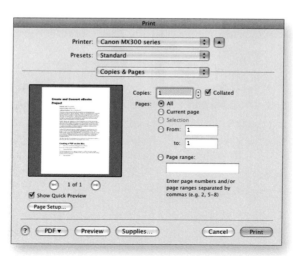

3. Click and hold the PDF button in the bottom-left corner to display the PDF menu (**Figure 3.11**).

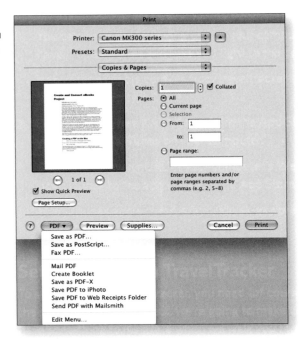

Figure 3.11 The PDF button displays a menu that includes a Save as PDF item.

4. Choose Save as PDF.

   A Save as dialog appears, letting you name your PDF and specify where it should be saved.

Now you have your PDF, and if you want to, you can drag it into iTunes for syncing to your iPad.

 Microsoft doesn't provide built-in PDF generation in its OS, but Windows users do have a free solution: CutePDF Writer (www.cutepdf.com/ products/cutepdf/writer.asp). You'll need to install the free Ghostscript package; the Web page has links and instructions.

Convert existing e-books

During the era of the word-processor and spreadsheet wars, every software developer employed proprietary formats to store its data on disk.

Users wanted conversion capabilities so that they weren't locked into a single provider and so that they could reuse content in other applications. Open formats, such as RTF for word processing and DIF or SYLK for spreadsheets, provided a least-common-denominator conduit between applications.

When e-book readers started appearing, the Tower of Babel re-emerged, with every producer of a reader employing its own proprietary (and occasionally secretive) document format. To name just a few, Amazon's Kindle uses AZW; Microsoft Reader uses LIT; Palm's eReader uses PDB; Mobipocket uses PRC and/or MOBI; the Nook, iPad, and several others use EPUB; and Sony Reader uses LRF. There are still more formats, and some devices can read formats other than their preferred format.

In the e-book world, the open formats are ePub and PDF (unless you consider HTML, plain text, and RTF to be e-book formats). ePub is a standard developed by the International Digital Publishing Federation. All these formats either include digital rights management (DRM) functionality or support extensions for DRM capability.

For more info on DRM, see the "Copy Protection and DRM" sidebar in the Streaming Your Video Project later in this chapter.

Adobe employs ADEPT, a proprietary DRM capability in ePub documents. We're amused by the name of an open framework that has been reverse-engineered to circumvent this DRM: INEPT.

In our opinion, the current best-of-breed format converter is an open-source project named calibre (www.calibre-e-book.com). The calibre software is free and available for Linux, Mac, and Windows, with only cosmetic variations appropriate to the platform differentiating the versions. If calibre were only a conversion utility, it would be great, but the application offers so much more. It also searches the Internet for metadata (publisher, publication date, genre, and so on), fetches newspaper and magazine content (such as articles from *The Economist*), and provides a viewer to let you read your e-books on your computer, to name just a few of its capabilities. The thrust of this project, though, is converting your e-books to ePub format for reading on your iPad, so let's get started.

E-Books for Your iPad

The Kindle lit the fuse, but the iPad lit up the sky. E-books have suddenly become the hottest topic in publishing, and almost every publisher now has an e-book publishing plan.

You can get a variety of e-book reader apps for your iPad, including Apple's iBooks, Amazon's Kindle, Barnes & Noble's eReader, and Lexcycle's Stanza. Here's the rub: Except for Stanza, each app is tied to one specific online bookstore, and you have to use that app to read your purchases from that store on your iPad.

Here's where and how to get books for the preceding e-book readers:

- **iBooks.** Tap the app's Store button to browse the integrated iBookstore. For iBookstore purchases, you use your iTunes Store account (which contains your credit-card information). Purchased books download to iBooks immediately. If you delete a book on your iPad (or on an iPhone that has the iBooks app), you can download it again for free.

- **Kindle.** Tap the Shop In Kindle Store button in the Kindle app, and it opens Safari on your iPad. When you purchase a book from the Kindle Store, it's sent immediately to the Kindle-compatible device you specify. (First, of course, you must register each of your Kindle-compatible devices with the Amazon Web site.) If you have a Kindle Reader app on your PC, on your iPhone, and on your iPad, for example, you choose which device gets the book first. Don't worry, though: You can download purchased Kindle books to the other devices from within the app on each of those devices. Also, you can always download deleted books again.

- **eReader.** This app acts like a hybrid of iBooks and Kindle. You tap Add Books to browse the bookstore from within the eReader app, but when you select a book, eReader opens Safari for you to make your book purchase. When you purchase a book, it appears immediately in your eReader library. Note, however, that it isn't *really* there until you actually tap the book's cover to open it, at which time it downloads to your iPad. Barnes & Noble also provides a book-lending feature, so you can lend a book to a friend or family member for a limited time.

- **Stanza.** Like eReader, Stanza lets you browse available books. Unlike the other apps, however, it provides access to several online bookstores. Because Stanza works with different bookstores, the process of getting a purchased book into the app can vary from store to store. You can browse the Fictionwise bookstore from within Stanza, for example, but when you tap a book to buy it, Stanza opens Safari for you to make your purchase. When the book is purchased, you can download it from within the Stanza app and unlock it (with your credit-card number) so that you can read it.

Stanza is the only e-book app that uses the file-sharing feature in iTunes, so you can drag books between your iPad and your computer. You can also share books between your computer and iPad wirelessly. Notably, Stanza is compatible with a much wider range of e-book formats than any of the other e-book apps. See the last task in this project for more details on using Stanza.

Getting and setting up calibre:

1. Navigate to www.calibre-e-book.com/download.
2. Select your operating system (**Figure 3.12**).

Figure 3.12 Click the icon for your OS to download the correct version.

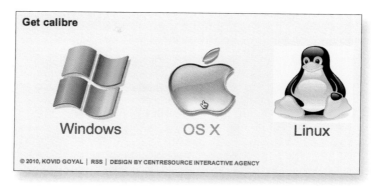

The site takes you to the download page for your OS (**Figure 3.13**).

Figure 3.13 Read the directions, and click the Download button.

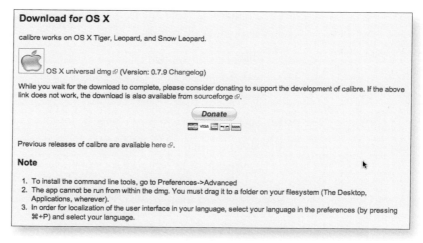

3. Click the download link (in our case, OS X universal dmg).

    Your browser starts the download. Depending on your OS and your settings, you may be asked what to do with the download (open it, save it to disk, and so on).

4. If the software doesn't install automatically, run the installer or (in Mac OS X) drag the calibre icon to your Applications folder.

5. Launch calibre.

   The welcome wizard greets you (**Figure 3.14**).

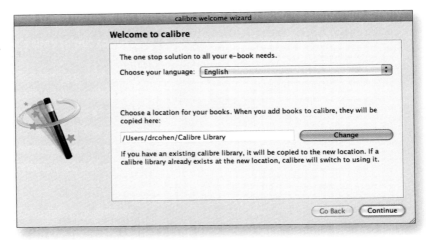

Figure 3.14 The welcome wizard walks you through calibre's setup process.

6. Choose your language from the pop-up menu (unless the default is what you want).

7. (Optional) If you don't like the default storage location offered for your books, click the Change button, and specify a new location in the Open dialog that appears.

8. Click Continue when you're ready for the next step.

9. Select the manufacturer of your e-book reader and the device you use (**Figure 3.15**); then click Continue.

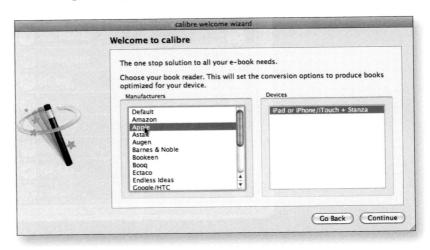

Figure 3.15 Let calibre know that Apple manufactures your e-book reader.

Chapter 3: Music, Books, and Movies on the iPad   **171**

If you use Stanza on your iPad (yes, we know that the dialog shown in Figure 3.15 mentions only the iPhone and iPod touch), you can turn on the content server and—so long as calibre is running and you're in range of your wireless network—access your computer-based book collection directly from your iPad. We provide a Stanza task later in this project and describe the Content Server there.

A congratulatory message appears, offering you access to tutorial videos and an online user manual. If you want to become a calibre power user, check them out. The folks behind calibre have done a superb job of demonstrating and documenting this (amazingly) free product's features and functionality.

10. Click Done.

calibre starts running, presenting a window that looks like **Figure 3.16**. The software starts you out with *Calibre Quick Start Guide*.

Figure 3.16 calibre is ready for use.

Assuming that you have some e-books available in formats other than ePub, you're ready to start adding them to your calibre library and converting them to ePub for your iPad, as we show you how to do in the following task. For this task, we're going to add Tess Gerritsen's mystery *Vanish* (the fifth book in her Rizzoli & Isles series, which is now seeing life on the TNT cable network). Our copy is in LIT format.

 If you don't have any non-ePub books, you can acquire them (sometimes for free) at several Internet locations. For the sci-fi fans among you, the Baen Free Library (www.baen.com/library/defaultTitles.htm) is a wonderful resource, even though you can actually specify the format you want, which removes the need to convert.

Adding and converting books:

1. Click the Add Books button at the far-left end of the calibre toolbar.

2. In your OS's Open dialog, navigate to and select a book; then click Open.

   The book is added to calibre's library list (**Figure 3.17**).

Figure 3.17 The book you selected is now in your calibre library.

tip — **If metadata is missing (as Publisher is in this example), click the arrowhead next to Edit Metadata, and choose Download metadata and covers from the menu that appears.**

3. Click the Convert Books button.

   The conversion dialog appears (**Figure 3.18**).

Figure 3.18 The conversion dialog has multiple panes; the Metadata pane is selected here.

4. Make any changes you desire in the metadata.
5. Click Page Setup in the list on the left side of the page.
6. Choose iPad in the Output Profile list.
7. Click OK.

   calibre performs the conversion. You can see the progress indicator spinning away in the bottom-right corner of the window, with a Jobs label (no relation to Steve).

8. Click the Save to Disk button (refer to Figure 3.17).
9. In the Open/Choose dialog that appears, select a destination directory.

   In the Finder (Mac) or Windows Explorer, calibre opens a window showing the directory you specified.

10. Open the folder for the author and then open the title folder.

11. If iTunes isn't running, launch it.

12. Drag the EPUB file from the title's folder into the Library section of the iTunes Source list.

    Your book is now in the Books section of your iTunes Library, ready for your next iPad sync (**Figure 3.19**).

Figure 3.19 Your converted title is in iTunes, ready to sync to your iPad.

 Alternatively, if you don't have iTunes running, or if you aren't into dragging and dropping, click the arrow next to the Send to Device icon and choose Connect to iTunes from the menu that appears.

iBooks isn't the only e-book reader available in the App Store. In fact, it isn't even the only free e-reader. Amazon distributes a free Kindle Reader, for example, and the open-source Stanza eReader is also available (see the "E-Books for Your iPad" sidebar earlier in this project). Stanza in particular plays very nicely with calibre, and if you've completed the task in this section, you've set up calibre with the iPad as your e-reader, so calibre is all ready to beam content to Stanza on your iPad.

Chapter 3: Music, Books, and Movies on the iPad   **175**

Sending content to Stanza:

1. Open calibre's Preferences dialog by clicking the rightmost button on the toolbar (refer to Figure 3.17 earlier in this project).

2. In the list on the left side of the dialog box, select Content Server.

3. If the Server isn't running, click the Start Server button to start it.

 We select the Run server automatically on startup check box to avoid this step.

4. On your iPad, launch the Stanza app by tapping its icon.

5. Tap Get Books at the bottom of the Stanza screen.

6. Tap the Shared tab at the top of the screen.

    A screen similar to the one shown in **Figure 3.20** appears, listing any sources or servers set up to share books to Stanza. In our example, the source is the calibre application on Dennis's MacBook.

Figure 3.20 Tap the calibre server's entry in Stanza on your iPad.

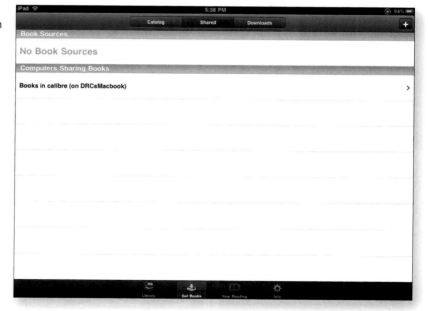

7. Tap the server from which you want to receive content.

   A list appears, letting you select how you want to sort the books: By Newest, Title, Ratings, Publishers, Series, Authors, News, or Tags. (You can read about the last two options in calibre's user guide.)

8. Choose a sort option.

   We're choosing By Series to see a list of series that are available and how many books are available in each series—for this example, just the one we created with the conversion of *Vanish* in "Adding and converting books" earlier in this project.

9. Tap the series you want.

   A display of the titles available in that series appears (**Figure 3.21**).

Figure 3.21 calibre displays the titles in a series.

10. Tap the title you want.

    A screen displaying your book's cover image appears, with a Download button in the top-right corner (**Figure 3.22**).

Figure 3.22 You can download the displayed title from this screen.

11. Tap the Download button.

    A confirmation dialog appears in the middle of your iPad's display.

12. Tap Download in the dialog to make that happen.

    When the download is complete (usually, very quickly), the Download button changes to Read Now.

13. Tap Read Now if you want to go directly to the book, or backtrack to add more books to your iPad's Stanza library.

At least for now, we find reading books in Stanza to be more convenient with the iPad in portrait orientation, whereas the two-page landscape display in iBooks is more to our liking. Your mileage may vary.

Movie and TV-Show Syncing Project

Difficulty level: Easy

Software needed: iTunes

Additional hardware: None

Television. The boob tube. The vast wasteland. What Shakespeare might have called "an expense of spirit in a waste of shame."

You know you love it. We certainly do. From the ghostly black-and-white haunted aquariums of the early 1950s to the giant flat-screen high-definition full-color sets of the 21st century, we can't get enough of TV. So the first time we saw video playing on an iPad, we moaned, "Gimme some of that. Now!"

In this project, you'll see how to get movies and TV shows syncing between your iPad and your computer so that no matter where you are, you can get the video fix you crave.

 The Music Syncing Project earlier in this chapter describes a way for you to manage music and videos manually. If you've set this option on your iPad in iTunes' Summary tab (**Figure 3.23**), you can't sync video between your iPad and computer automatically. This project assumes that you *don't* have that option turned on.

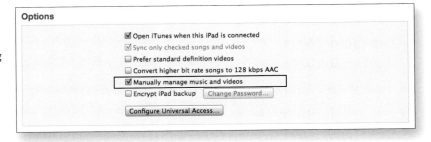

Figure 3.23 If you want to sync movies and TV shows, don't check this iPad syncing option.

 When you make any changes in the iTunes syncing tabs, the changes don't take effect immediately, so you can try out different settings. When you click something else in the iTunes Source list, iTunes asks whether you want to apply your changes. You can also click Apply in the bottom-right corner of the iTunes window to apply your changes immediately, or you can click Cancel to set all your unapplied syncing changes in all the syncing tabs back to the way they were (**Figure 3.24**).

Figure 3.24 Use the two buttons in the bottom-right corner of the iTunes window to apply syncing changes immediately or to cancel changes and revert to previous settings.

Sync movies

Whether you get your movies from the iTunes Store or somewhere else (and we discuss some of the "somewhere else" options elsewhere in this chapter), getting them from your iTunes Library to your iPad and back again isn't rocket science, even though the technology underlying it may be.

You can sync movies by the following criteria:

- All movies

- Unwatched movies (*unwatched* movies being those that you haven't watched all the way through to the end)

- The most or least recently obtained unwatched movies

- Selected movies

- Selected playlists of movies

 Yes, you can create playlists in iTunes that contain movies. This feature is especially useful for short movies, such as the ones you make yourself with iMovie or some other movie-creation application.

Some of these criteria aren't mutually exclusive. You can sync five recent unwatched movies as well as additional selected movies and playlists of movies, for example.

In this section of the project, you get to experiment with various movie-syncing settings to see how they work. As noted earlier, the changes don't take effect without your say-so.

Turning on movie syncing:

1. Connect your iPad to your computer.
2. Launch iTunes.
3. Select your iPad in the iTunes Source list, below the Devices heading.
4. In the main pane of the iTunes window, click the Movies tab.
5. At the top of the Movies tab, select Sync Movies (**Figure 3.25**).

Figure 3.25 This check box is the master key to iPad movie syncing.

If **Manually manage music and videos** is checked in the Options section of the Summary tab (see the intro section of this project), turning on movie syncing unchecks that option. When that happens, the syncing options in the Music and TV Shows tabs take effect, so you need to look in those tabs as well and set some syncing options.

When movie syncing is enabled, you have access to the other controls and lists in the Movies syncing tab. First, you see how to turn on syncing for all movies.

Syncing all movies:

1. Below the Sync Movies heading in the Movies tab, click the Automatically include *x* movies check box.
2. Choose the option titled all from the pop-up menu in the middle of the preceding option's name (**Figure 3.26**).

   The rest of the options in the Movies tab vanish at this point; with all movies set to be synced, you don't need them.

Figure 3.26 When you sync all movies, the other options in the Movies syncing tab vanish.

Keep in mind that movies take up a lot of room. If you sync all movies, iTunes copies only as many as can fit on your iPad. You may find that you don't have room for many apps, books, or songs if you fill your iPad with movies.

In most cases, you probably *don't* want to sync all your movies, just as you wouldn't pack every piece of clothing you own when you go on a trip. You usually want your iPad to have the newest movies in your collection or the ones that you haven't yet viewed. In the following task, we show you how to make it so.

Syncing new or old movies:

1. Below the Sync Movies check box in the Movies tab, check the Automatically include *x* movies option, if it isn't already checked.

2. From the pop-up menu in the middle of the preceding option's name, choose any option other than all or all unwatched (**Figure 3.27**).

Figure 3.27 Your choices for syncing movies automatically.

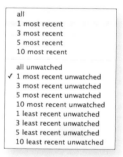

When you make any choice other than all or all unwatched, all the lists of movies and playlists in the Movies syncing tab become active.

 iTunes uses the date when each movie was added to the iTunes Library—not the movie's release date—to figure out which movies are the most or least recent. From the pop-up menu, you can choose to sync automatically one or more of the most recent watched movies; you can choose one or more of the most recent unwatched movies; or you can choose one or more of the least recent unwatched movies. For some reason, however, you can't choose to sync the least recent watched movies: those, you have to select manually (**Figure 3.28** on the next page).

Figure 3.28 You can manually select additional movies when you choose a specific number of recent or unwatched movies to sync automatically; that's why the lists of movies and playlists are available below the syncing option.

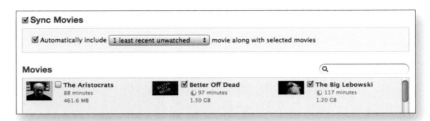

Speaking of including selected movies and playlists when you sync, that task is coming up next.

Syncing selected movies and movie playlists:

1. Below the Sync Movies check box in the Movies tab, do one of the following things:

   - Clear the Automatically include *x* movies check box.

   - From the pop-up menu in the middle of the Automatically include option's name, choose an item other than all.

   The lists below the option's name become available (**Figure 3.29**).

Figure 3.29 You can pick movies individually and choose playlists of movies.

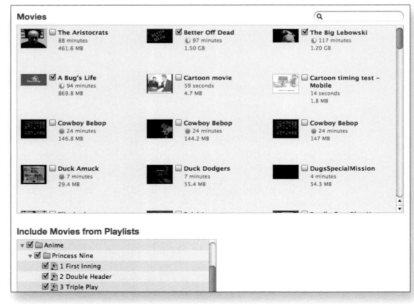

2. Click to select the movies you want to include along with your choices from step 1.

 If you choose any of the unwatched items from the Automatically include pop-up menu in step 1, you can select only additional movies that don't match the unwatched items you've chosen. The movies that match your choice are automatically checked in the Movies list and can't be unchecked.

3. In the Include Movies from Playlists list, click to select the playlists you want to sync.

Sync TV-show episodes

The videos in your iTunes Library that are categorized as TV shows have two special items of information associated with them that affect syncing: the name of the TV series and the episode of that series. You can use these two pieces of information to specify which TV-show episodes get synced.

The choices you have for syncing TV-show episodes are similar to those for movies:

- All episodes
- Unwatched episodes
- The most or least recently obtained unwatched episodes
- Selected episodes
- Selected playlists that contain TV-show episodes

Because episodes "belong" to TV shows, you can specify whether the unwatched and recent criteria apply to all TV shows or only to selected TV shows. If you're an avid fan of Jerry Van Dyke, for example, you can choose to sync only the five oldest unwatched episodes of *My Mother the Car* and no others.

As with movies, in this part of the project you get to experiment with various episode syncing settings to see how they work. Also as with movies, any changes you make in TV-show syncing don't take effect without your approval.

Turning on TV-show syncing:

1. Connect your iPad to your computer.
2. Launch iTunes.
3. Select your iPad in the iTunes Source list, below the Devices heading.
4. In the main pane of the iTunes window, click the TV Shows tab.
5. At the top of the TV Shows tab, select Sync TV Shows (**Figure 3.30**).

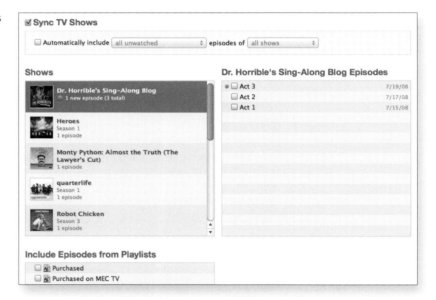

Figure 3.30 What lies beneath the TV Shows tab.

Because TV shows consist of many episodes, only some of which you may have seen, the options for syncing new and old episodes are more flexible than they are for movies. Let's take a look.

Syncing new or old episodes:

1. Below the Sync TV Shows check box in the TV Shows tab, check the Automatically include *x* episodes of *x* option, if it isn't already checked.
2. From the first pop-up menu in the Automatically include option, choose any item other than all.

3. From the second pop-up menu in the Automatically include option, choose all shows.

   (You'll deal with selected TV shows a little later in this project.)

4. Select a TV show in the Shows list.

   The Episodes list to the right shows which, if any, of the show's episodes are set to sync automatically (**Figure 3.31**). You can click the check boxes next to other episodes in a show's Episodes list to include those episodes in the sync.

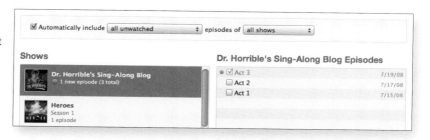

Figure 3.31 You can choose other episodes in addition to those that sync automatically.

> **tip** In addition to the episodes that fall within the syncing criteria you set, you can put TV episodes in playlists and select those playlists in the Include Episodes from Playlists list at the bottom of the TV Shows syncing tab (refer to Figure 3.30 earlier in this project).

Now that you've seen your choices for choosing episodes by how long you've had them and whether you've watched them, in the next task we show you how to specify the episodes that get synced.

Syncing episodes of selected TV shows:

1. Below the Sync TV Shows check box in the TV Shows tab, click the Automatically include option, if it isn't already checked.

2. From the second pop-up menu in the preceding option (the choice in the first pop-up menu doesn't matter), choose selected shows.

   Check boxes appear beside each show's name in the Shows list.

3. In the Shows list, click a show's check box to select it.

Only the episodes from the selected TV shows that match the criteria in the first pop-up menu sync automatically. As shown in **Figure 3.32**, however, you can select other shows in the Shows list and sync individual episodes of them as well.

Figure 3.32 You can sync episodes from shows other than the ones you've chosen to sync automatically.

Finally, if you happen to be a completist when it comes to TV shows, the last task in this project shows you how to sync all your TV shows.

Syncing all TV shows:

1. Below the Sync TV Shows check box in the TV Shows tab, click the Automatically include option, if it isn't already checked.

2. From the first pop-up menu in the Automatically include option, choose all.

3. From the second pop-up menu in the Automatically include option, choose all shows.

   All the lists in the TV Shows syncing tab vanish. After all, you don't need them if you're syncing everything.

4. In the bottom-right corner of the iTunes window, click Cancel.

   In this project, you've been playing around with a lot of settings to see what they do. Unless you want them to take effect—and chances are excellent that you don't—it's best to cancel and then plan how you *really* want to sync your TV shows and movies. After all, now you know how to do it; that's what this project has been about.

   Happy viewing! But first, stay tuned for these important messages….

What About Music Videos?

In addition to TV shows and movies, you may have music videos in your iTunes Library—videos that were included with albums you purchased or that you purchased individually. Music videos sync according to the same criteria you set for the songs in your Music library—*if* you set the right music syncing option.

In the Music syncing tab, in the first group of options, select the Include music videos option. That's it. Now when you sync your music, your music videos come along for the ride. You'll find them on your iPad in the Video app's Music Videos tab.

Moving Movie Rentals

You can't sync movies that you've rented from the iTunes Store on your computer. You can *move* rentals from your computer to your iPad and back, but the rental is always on only one device at a time.

In addition, if you rent a movie from the iTunes Store on your iPad, it stays on your iPad; you can't move it to your computer or any other device. (Similarly, if you rent a movie via Apple TV, you can't move it from Apple TV to any other device.)

Streaming Internet Video Project

Difficulty level: Moderate

Software needed: ABC Player (free at the App Store)

Additional hardware: None

Everywhere we turn, we find people catching their TV shows at times other than the scheduled broadcast times and, frequently, on devices other than a TV set or home entertainment center. Battling viewership loss to cable, and in an attempt to recoup ad revenue lost to cable and other competition, network television has begun to embrace alternative delivery systems—in particular, the Internet and iTunes.

Time and venue shifting have become so pervasive that Nielsen Media Research, in an attempt to stay relevant, purchased NetRatings to measure the demographics of the rapidly growing Internet viewing population. Nielsen also factors iTunes and YouTube viewing into its ratings.

Get the software

Although all the networks stream their shows via their Web sites, and though Safari on your iPad provides a competent content conduit, ABC has taken things a step further, providing a dedicated iPad app called ABC Player to stream its shows to our handheld devices.

This app offers a viewing experience that's tailored to the iPad rather than the lowest-common-denominator Web browser interface—as you see in **Figure 3.33** and **Figure 3.34**, which were taken seconds apart on the same day.

Figure 3.33 ABC's Web interface.

Figure 3.34 ABC Player's interface.

Obtaining ABC Player:

1. Tap the App Store icon on your iPad's home screen.

2. In the Search field in the top-right corner of the App Store's home screen, type **ABC**; then tap ABC Player in the list that appears.

3. Tap the ABC Player entry (which should appear in the top-left corner of the iPad Apps section).

   The ABC Player product page appears.

4. Tap the Free button below the icon in the top-left corner.

   Free flips over and becomes Install.

5. Tap Install.

6. When you're asked to enter your Apple ID and password, do so.

   You're back at the home screen, and ABC Player is downloading.

You're ready to start enjoying ABC's network TV shows. (At least, ABC hopes that you'll enjoy them.) Tap the ABC Player icon to enter the world of ABC prime-time (and more) television.

Use the ABC Player

When you have ABC Player running, you find five buttons along the bottom of the screen: Featured, Schedule, All Shows, Me, and Info.

Viewing ABC's featured favorites:

1. If it isn't already selected (which it is by default), tap the Featured button at the bottom of the ABC Player screen.

    You see a large thumbnail at the top of the screen, displaying a show that ABC is—wait for it—featuring (**Figure 3.35**). This display slides to the left about every 5 seconds to bring a new show's thumbnail into view.

Figure 3.35 ABC's featured shows and episodes on a summer day in 2010. Your mileage may vary.

2. To speed the movement of the large thumbnail display or to make it go in the opposite direction, touch the thumbnail and then flick your finger in the desired direction.

3. To watch a recent episode, do one of the following things:

    - To watch the most recent episode of a series during its regular season, tap the Watch Latest Episode button to—yes—watch the latest episode of that show.

- To watch episodes of a series that's between seasons, such as *Desperate Housewives* in Figure 3.35, the button will say Catch Up on Season *n* (6, in the case of the ladies on Wisteria Lane).

4. To view a specific episode of one of the shows displayed below the scrolling thumbnail, tap that episode's thumbnail.

5. To modify what's being displayed in the various thumbnails, tap Most Popular, Most Recent, or Staff Picks.

Seeing what's on the schedule:

1. To check out the network schedule, tap the Schedule button at the bottom of the ABC Player screen.

   You see a screen resembling the one shown in **Figure 3.36**.

Figure 3.36 Pick an ABC episode from the weekly calendar display.

2. Tap the desired day of the week to find that day's shows in their ABC time slots.

   As you can see in Figure 3.36, some shows aren't available for viewing on the iPad—usually, local broadcasts, sporting events, and the like. Also, shows that are available in the iTunes Store are so marked, with a button you can tap to buy them.

Checking it all:

1. Tap the All Shows button at the bottom of the ABC Player screen.

   You see a grid of all available ABC shows (**Figure 3.37**).

Figure 3.37 Find the show you want in ABC's All Shows grid.

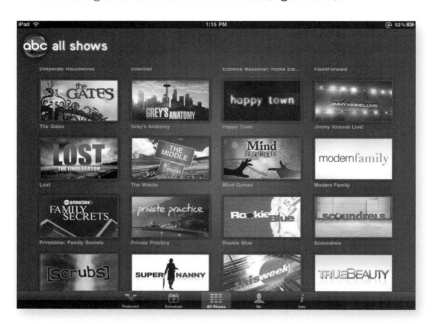

2. Tap a show's thumbnail to see a pop-up list of episodes (**Figure 3.38**).

Figure 3.38 Tap the show to see a list of available episodes.

3. Tap the desired episode to start viewing.

Reviewing your viewing history:

1. Tap the Me button at the bottom of the ABC Player screen.

   You see a grid of thumbnails representing the episodes you've seen so far (**Figure 3.39**). At the top of each thumbnail is a timeline, with a down-pointing arrow showing where you left off watching the episode. Near the top-right corner of each thumbnail is a small x that you can tap to remove the episode from your history.

Figure 3.39 Check here in case you don't remember what you've already watched (or if you like reruns).

2. (Optional) If you see an episode you want to watch again or want to pick up watching an episode where you left off, tap its thumbnail.

 This feature is great if you regularly watch a few TV series and have seen multiple episodes of each one. It's very easy to lose track of which episodes you've viewed and which you haven't. Thanks to the Me button, you don't have to remember multiple episode names in multiple series.

Giving ABC feedback:

1. Tap the Info button at the bottom of the ABC Player screen.

   A feedback form appears (**Figure 3.40**). This form is a lot less colorful than the player's other pages, but you hold sway here.

Figure 3.40 Tell ABC what you think, or make suggestions.

2. Fill out the form to tell ABC what you consider to be good or bad, or to suggest features you'd like to see in future versions.

3. When you've had your say, click the Send Feedback button.

View the video stream

After you tap an episode in the Featured, Schedule, All Shows, or Me screen (well, some people like to watch shows multiple times, so if you're not one of them, you don't need to tap an episode in your history), ABC Player starts loading the episode.

During the loading process, the player displays a placard stating that the following episode is being presented with limited commercials. That may be true, especially if you're defining *limited* by counting the distinct advertising spots, because you generally get the same two or three commercials repeated ad nauseam through the show. It also may be true if your tolerance for interruptions is greater than ours. We count four interruptions per hourlong episode in addition to the lead-in commercials, with two commercials per interruption. (Actually, an "hourlong" episode viewed in ABC Player is typically 42 to 43 minutes long, not counting the commercial time.)

When the introductory commercial break is over, the episode starts to play.

Controlling video playback:

1. While you're viewing an episode in ABC Player, tap a video to display viewing controls at the bottom of the screen (**Figure 3.41**).

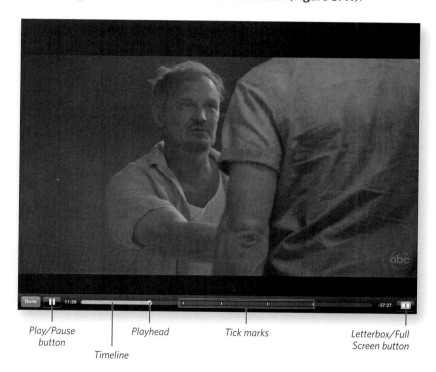

Figure 3.41 Tap an episode to control its playback.

2. Do any of the following things:

    - Tap the blue Done button to go back to the show's episode screen.
    - Tap the Pause/Play button to toggle between pause and play.

- Tap the Letterbox/Full Screen button to control the video's aspect ratio.

- Drag the playhead in the timeline to move to a different section of the episode.

Be aware, however, that if you drag over one of the tick marks (each of which denotes a "limited commercial interruption"), the playhead lands on the commercial rather than on the particular point you chose.

When a commercial is playing, the viewing controls are inaccessible. You can't pause playback, much less fast-forward or rewind. In fact, you can't even end playback and return to the selection page. This situation is particularly annoying when you accidentally tap one of the thumbnails, because every video starts with one of these uninterruptible commercials. Thus, you have to sit through the commercial until the controls are available again, or press the iPad's Home button and then relaunch the **ABC Player** app.

Streaming Your Video Project

Difficulty level: Intermediate

Software needed: Air Video ($2.99 from the App Store) and Air Video Server (free)

Additional hardware: Mac or Windows computer

We love TV and movies, and as we write in the **Converting Video Project** later in this chapter, the iPad is a fantastic platform on which to watch our video. Unfortunately, the iPad doesn't support most of the formats in which video is commonly distributed. It likes only MPEG-4 (MP4 or M4V)—no MPEG-1 or MPEG-2 (used for VCDs [video compact discs] and DVDs); no QuickTime; no AVI, DivX, or Xvid; no Matroska (MKV); and no Adobe Flash, just to name some common formats that you'd need to transcode to play on your iPad. Every one of these transcoding operations involves a diminution of quality resulting from recompression to the new format.

Add to that fact these considerations: Your iPad's storage space is fairly limited, and video takes a lot of space. Your desktop (or laptop) computer, on the other hand, usually has a pretty large hard drive, and you can extend that space with additional drives. Wouldn't it be nice to use your iPad to view the video stored on your computer without having to sync it?

Thanks to InMethod's Air Video products, you can do just that. Read on.

Get Air Video

You need two pieces of InMethod software:

- **Air Video or Air Video Free.** You can get Air Video at www.inmethod.com or from the App Store. If Air Video's $2.99 price tag puts you off, or if you just feel compelled to try before you buy, you can obtain Air Video Free (available from the same sources), which limits the number of videos accessible in each folder.

- **Air Video Server.** You also need the free server software, available at www.inmethod.com. Air Video Server requires Mac OS X 10.5 (Leopard) or later or Windows XP Service Pack 3 or later (the same Windows software required for iPad support).

Air Video streams virtually any video on your computer's hard drive, transcoding it as necessary. The lone exception is copy-protected video, which you usually purchase or rent from the iTunes Store (see the sidebar on the next page).

> **tip** **Although Air Video will transcode on the fly, which is the way most users employ it, you can perform the conversions before streaming the content. If you have a slower network (or a computer slower than a Core 2 Duo), you may want to perform the up-front conversion.**

Copy Protection and DRM

Rather than call it *copy-protected,* a term that has serious negative connotations in the public mind, the industry and lawyers refer to this type of content as *digital rights management* (DRM) content.

Copy protection and DRM are the same thing when you're dealing with digital material. Because the DRM content you obtain from the iTunes Store is already in an iPad-compatible format, and you can sync it to your iPad, why can't Air Video stream it? Well, that situation arises because the DRM license that the lawyers impose on Apple precludes streaming of DRM content. Therefore, if you want to watch video from the iTunes Store on your iPad, that's about the only video you'll need to sync.

Obtaining the Air Video software:

1. Point your Web browser to www.inmethod.com.

2. Click the link labeled Get Air Video from the iTunes AppStore, or click the link for the free version.

3. In the App Store, follow the familiar steps to complete your acquisition and download the app to your iTunes Library.

4. Sync your iPad so that the software is installed on your iPad.

You can combine steps 2-4 by connecting to the App Store on your iPad and making the purchase there. Remember that Air Video won't be backed up to your computer until your next sync.

5. Back on the InMethod Web site, click the graphic for your operating system (the blue Apple logo if you're using a Mac or the Windows logo if that's your platform) to download the Air Video Server software for your computer.

    The Server software for your platform downloads and installs itself on your computer.

note If the Server software doesn't install automatically when you download it, run the installer (Windows) or drag the Server application's icon to your Applications folder (Mac).

Introduce your iPad to Air Video Server

Now that you have Air Video on your iPad and Air Video Server on your computer, it's time to open the lines of communication.

Setting up Air Video Server:

1. Launch Air Video Server.

   You should see the dialog shown in **Figure 3.42**.

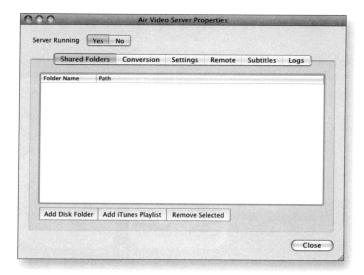

Figure 3.42 The Air Video Server Properties dialog.

2. If you want to specify a folder (such as your Movies folder on a Mac or your MyMovies folder in Windows), click the Add Disk Folder button, and navigate in the Open dialog to select your folder.

   Now your specified folder appears in the dialog's list box, as shown in **Figure 3.43** on the next page.

Figure 3.43 Your first entry in the folders that Air Video can access.

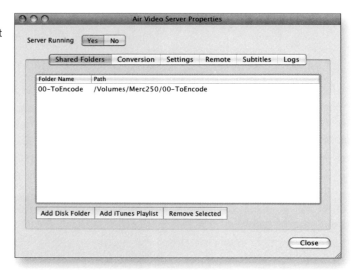

3. Repeat step 2 for any additional folders you want to make available.

   Note that selecting a folder also selects all subordinate folders.

4. To add iTunes playlists to the list of streamable locations, click the Add iTunes Playlist button.

   The iTunes Playlists dialog opens (**Figure 3.44**).

 As we mention earlier in this project, copy-protected content from the iTunes Store won't stream. iTunes playlists are handy for your home videos, iTunes U courseware, video podcasts, and other content that lacks DRM.

Figure 3.44 Select any iTunes playlists you want to access.

5. Select the iTunes playlists (or categories) you want to add, and click Close.

   You return to the Air Video Server Properties dialog.

6. Make sure that the Server Running switch at the top of the window is set to Yes, as shown in **Figure 3.45**.

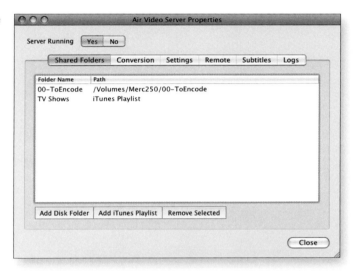

Figure 3.45 Make sure that the Server is turned on.

Air Video Server runs as a background process. No icon for it appears in either the Mac OS X Dock or the Windows taskbar. If you close the window, the application continues to run, and a dialog (**Figure 3.46**) informs you that you can still access it via an icon in the application's menu bar (Mac) or system tray (Windows).

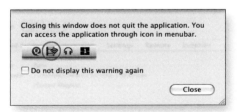

Figure 3.46 Use the menu-bar or system-tray icon to access the faceless Server application.

You're good to go and should find your specified content available in the Air Server app on your iPad (**Figure 3.47** on the next page).

Figure 3.47 Your selected folders are now available on your iPad.

Setting Additional Server Preferences

In the Air Video Server Properties dialog (refer to Figure 3.42), you can click the other tabs to set various preferences that launch the application automatically at login, require a password, or set a custom port (Settings tab).

You can also instruct the Server to honor Internet access (Remote tab), but this functionality requires that your router support UPnP (Universal Plug and Play) or NAT-PMP (Network Address Translation-Port Mapping Protocol). When you select the Enable Access from Internet check box, a server PIN is displayed. Make certain that the Automatically Map Port check box is selected. Now you can access Air Video Server from remote locations via Wi-Fi or (if Wi-Fi isn't available and you have a 3G iPad) via AT&T's 3G network.

Caution: 3G access is slower than Wi-Fi, so live conversion may be more problematic. Also, with AT&T's pricing, the amount of data involved can burn through your monthly 3G allotment fast, causing you to run up some hefty overage charges.

Play your content

Playing your content is simple, just as you'd expect on the iPad. Air Video's iPad user interface operates in landscape orientation, even when the iPad is vertical. We find this fact somewhat amusing, because Air Video started as an iPhone app.

At any rate, if you're using your iPad in portrait orientation, rotate it 90° to landscape or tilt your head 90° (if you don't mind being uncomfortable).

Playing a video:

1. Select the video you want to play, tunneling down through your folder hierarchy if necessary.

    The beginning of your video appears in the preview pane on the right side of the iPad screen (**Figure 3.48**).

Figure 3.48 The preview pane includes controls that start your video, convert your video, and manage the conversion queue.

2. Tap the Play with Live Conversion button if you're into immediate gratification, or tap the Convert button if you're a disciple of Job.

   For the purposes of this task, choose to be impatient.

   > **tip** **If you do elect to perform the conversions before you play the video, you can check the queued conversions and their status by tapping the Queue button in the top-right corner of the preview pane.**

   After a quick spin of the wait cursor, your video starts to play in the preview pane.

3. Tap the double-arrow button in the bottom-right corner of the preview pane (**Figure 3.49**) to have your video appear full-screen.

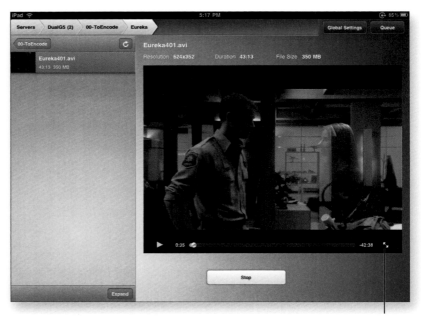

Figure 3.49 When video is playing, controls are available at the bottom of the preview pane.

Tap for full-screen video.

4. In the controller overlay (**Figure 3.50**), do any of the following:

    - Tap the 30-Second Rewind button to go backward in 30-second increments.

    - Tap the Play/Pause button to switch from playing to pausing, and vice versa.

    - Tap the Advance button to move to the next chapter marker (or the end of the video, if there are no remaining chapter markers).

    - Tap the Exit Full Screen button if you want to go back to the screen with the preview pane (refer to Figure 3.49).

    - Drag the scrubber in the timeline to move to any point in the video you want. The current time code (how far into the video you are) is shown at the left end of the timeline, and the remaining time in the video is displayed at the right end.

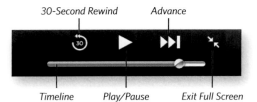

Figure 3.50 Video displayed in full-screen mode and the controls available in that mode.

More Convenience Features

Here are a few tips for added enjoyment and functionality:

- If the controls aren't visible, just tap within the video to make them appear.

- Double-tap the video in full-screen mode, and the video expands to fill the screen, even if that expansion results in cropping the sides of a widescreen show. You can do the same thing by tapping the double-arrow button above the timeline (refer to Figure 3.50), switching between letterbox and full-screen view.

- If you've added videos to or removed videos from the folder currently displayed in the list on the left side of the preview pane, tap the Refresh button above the top-right corner of the list (refer to Figure 3.48) to update the display.

Watching Television Project

Difficulty level: Moderate

Software needed: EyeTV 3 software, version 3.4 or later ($79.95), EyeTV app ($4.99)

Additional hardware: Intel Macintosh (required), EyeTV capture product or a compatible hardware solution, Turbo.264 HD recommended for streaming over 3G

For this project, only Mac users need apply.

Ever since Elgato Systems released its first EyeTV tuner back in the autumn of 2002, the EyeTV product line has been garnering awards and four- and five-star ratings. In our opinion, based on longtime personal use as well as on reading the product reviews, the quality of Elgato's software is a major reason—if not *the* major reason—for the high ratings and the awards.

Elgato is a rarity among companies that produce a varied line of hardware peripherals because its product line focuses on Macs and Mac users. Though some of the hardware (such as the EyeTV Hybrid) works well on a Windows 7 system with Windows Media Center, the software is Mac-only. The EyeTV software and hardware combination turns your Mac into a television set, a DVR, a video-editing station, and more. That *more* includes redirecting TV to another Mac, an iPhone, or an iPad.

Elgato also develops and maintains Toast Titanium for Sonic/Roxio. Toast is the leading CD/DVD/Blu-ray burning solution for the Macintosh, and it integrates beautifully with EyeTV for burning video to DVD. Because EyeTV also takes input from a VCR, it's a very handy way to transfer your old VHS (or Beta) tapes to DVD.

Get the software, hardware, and app

You can check out the Elgato hardware and software product line at www.elgato.com/elgato/na/mainmenu/products.en.html, or just go to www.elgato.com and click the Products link on the home page.

We aren't going to tell you how to hook up an EyeTV unit to your Mac. Elgato does that extremely well in its product documentation; besides, the instructions vary depending on which tuner you purchase. Similarly, we won't tell you how to install the EyeTV 3 software, which installs like any other Mac software.

After you connect the EyeTV hardware and install the EyeTV 3 software, launch that software and then follow the onscreen setup instructions for your tuner.

Next, visit the App Store. Click (Mac) or tap (iPad) your way to the Store, and purchase the EyeTV app (current price: $4.99).

If you purchase the app via iTunes on your Mac, sync your iPad so that the EyeTV app will be on your iPad, and if you purchase it on your iPad, sync anyway so that the backup copy will be available on your Mac.

When you have all the pieces in place, it's time to get them talking, which you do in the following task.

EyeTV's Transition from FireWire to USB

When Elgato introduced the first EyeTV units, every Mac came with one or more FireWire ports, and the only USB ports were USB 1.1. Additionally, digital camcorders connected via FireWire (IEEE-1394 or, as Sony called it, i.Link), making FireWire the perceived digital video bus of choice.

The EyeTV line moved from its original FireWire implementation (in the days when Macs sported USB 1.1 ports, which were far too slow for video transport) to USB when Apple transitioned its iPod line to USB 2.0. FireWire has higher sustained throughput than USB 2.0, but the difference is negligible.

Apple's decision not to include FireWire in some models (currently, the MacBook and MacBook Air) and to significantly reduce the number of FireWire ports in the rest of the product line proved to be serendipitous.

Getting everything linked:

1. Tap the EyeTV app to get it running on your iPad.

   After the short splash screen fades, you'll see the screen shown in **Figure 3.51**.

Figure 3.51 The EyeTV app tells you to turn sharing on in your Mac's EyeTV software.

2. In EyeTV 3 on your Mac, choose EyeTV > Preferences to open the Preferences dialog.

3. Click the Sharing button in the Preferences dialog's toolbar, and select the Share my EyeTV Archive check box (**Figure 3.52**) to let EyeTV on other Macs (or your iPad/iPhone) see this Mac's archive.

Figure 3.52 EyeTV's Sharing preferences.

The Look for shared EyeTV Archives option is handy if you have other EyeTV units running on other Macs on your network. (Dennis 'fesses up to having three going.)

Your EyeTV archive source now appears in the EyeTV app on your iPad.

4. Tap the archive source.

   You see the screen shown in **Figure 3.53**.

Figure 3.53
The EyeTV app says that you still have another preference to set on your Mac.

5. In the EyeTV Preferences pane on your Mac, select the check box titled Enable access from EyeTV for iPhone/iPad (**Figure 3.54**).

Figure 3.54 Select this check box to let your iPad access this EyeTV archive.

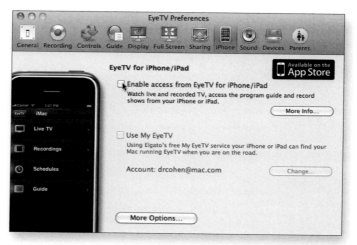

The Use My EyeTV option lets your iPad (or iPhone) locate your Mac's EyeTV software over the Internet.

6. Back on your iPad, click the bar that represents a Mac sharing its EyeTV archive.

   Now your iPad's EyeTV app has access to your Mac's EyeTV software, including live TV, its existing recordings, and the ability to schedule recordings (**Figure 3.55**).

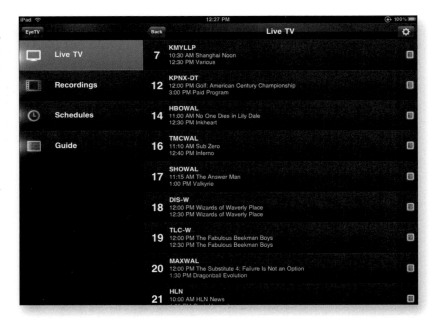

Figure 3.55 Your EyeTV app can access your EyeTV hardware, recordings, schedules, and Guide.

Navigating EyeTV:

You have several ways to get around in and use EyeTV:

- In the EyeTV app on your iPad, tap a show in the Live TV screen, which displays the current schedule.

  EyeTV connects to your Mac and starts receiving the requested broadcast as your Mac transcodes it for display on your iPad. (For more info about transcoding, see the last section of this chapter: the **Converting Video Project**.) **Figure 3.56** shows TV playing on an iPad.

Figure 3.56 Watch live TV on your iPad.

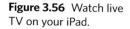 **tip** This process is where an Elgato Turbo.264 HD unit comes in handy. With the Turbo.264 in play, the conversion is much faster, and it offloads the processing from your Mac's CPU. The original raison d'etre for the Turbo.264 HD was to facilitate bringing AVCHD (the high-def version of the Advanced Video Codec, also known as H.264) content in from high-definition camcorders, but the conversion for transmission to iPads and iPhones is an excellent side benefit.

- Tap the little green button at the far-right end of a channel's entry to see the schedule for that channel (just like the Guide, described later in this section).

- In the top-left corner of the Live TV screen (refer to Figure 3.55) is a button labeled EyeTV. Tap it to return to the display of EyeTV units your iPad sees.

- Tapping the Back button above the list of available shows lets you choose broadcast channels or A/V input. (You can have a VCR, LaserDisc player, or other analog source connected to your EyeTV tuner and select it as your input source.)

- Tap the Actions button in the top-right corner of the Live TV screen (it looks like a gear) to present the screen shown in **Figure 3.57**. The sliders in this screen let you adjust throughput over various types of network connections. (Also, this screen mentions how the Turbo.264 HD helps improve streaming and picture quality; refer to the tip earlier in this section.)

Figure 3.57 Use the sliders to adjust network access speed.

- In the list on the left side of the Live TV screen, choose any of the following options:

    - Tap Recordings to see a list of the recordings in the Mac's EyeTV DVR archive.

    - Tap Schedules to see what recordings are scheduled.

    - Tap Guide to see what's currently playing or (by tapping a channel's entry) to see what's on tap for that channel and schedule recordings, as shown in **Figure 3.58**.

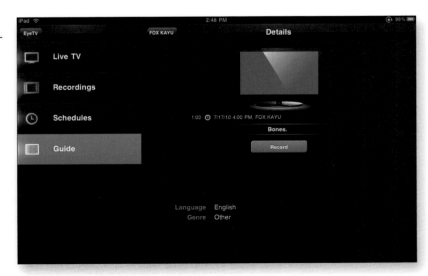

Figure 3.58 Tell your Mac's EyeTV to schedule a recording.

Converting Video Project

Difficulty level: Intermediate

Software needed: HandBrake (free), VLC (free), and iTunes

Additional hardware: Mac or Windows computer with DVD drive

The iPad is a wonderful video viewer—easily the best handheld viewer we've encountered. Even if the only video you watch on yours is content obtained through the iTunes Store, you have a rich (and possibly expensive) experience.

You can easily put your own videos on the iPad, though, or convert videos obtained from other sources. We're not lawyers, don't play them on TV, and don't profess to be them, so we won't get into the argument about whether *transcoding* (converting from one format to another) the content on DVDs you've purchased to play on your iPad is legal or actionable. We'll just note that an awful lot of people do it and that their arguments concerning fair use (see the nearby sidebar) are at least as compelling to consumers as the arguments advanced by the Motion Picture Association of America (MPAA) are to the trade. The judicial system still hasn't weighed in definitively, or even consistently, on the subject.

What Is Fair Use?

In U.S. copyright law, *fair use* is a doctrine allowing limited use of copyrighted material without the explicit permission of the copyright holder. Four criteria apply in copyright law, but the seminal point argued for private use rests on the decision in *Sony Corp v. Universal City Studios* in 1984 (aka "the Betamax case"), wherein the U.S. Supreme Court ruled that making individual copies for purposes of time shifting (recording a show to view at a later time, as we all do nowadays with DVRs) doesn't constitute copyright infringement.

The courts are currently considering whether this decision can be interpreted to include file sharing and conversion technologies such as transcoding to shift the viewing platform from a specific device (such as a DVD player) to another device (say, an iPod, iPad, or iPhone). Thus far, they haven't issued a definitive ruling on either side of the question.

In the meantime, HandBrake and similar technologies exist and are freely available.

Acquire HandBrake and VLC

The first thing you need to do is acquire a current version of HandBrake and a copy of VideoLAN Client, aka VLC.

HandBrake converts a large—and growing—collection of input video formats to the H.264 codec that your iPad wants. As an open-source project, HandBrake doesn't have a predetermined development and release schedule, so it's almost guaranteed that something new will be added between the time we're writing this project and the time you're reading it. Check out the HandBrake site to see what the current version is (at this writing, 0.9.4).

VLC is a very capable open-source, multiplatform audio and video player application—one that plays formats that aren't playable in iTunes or QuickTime without additional (and sometimes nonfree) plug-ins.

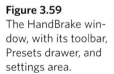

HandBrake requires VLC only for transcoding DVDs. If your intent is to convert AVI, DivX, Xvid, MKV, and other formats for use on your iPad, you really don't need VLC, but we still recommend it because it's a very useful viewer.

After you install both programs and launch HandBrake, you should see the window shown in **Figure 3.59** (or something quite similar).

Figure 3.59
The HandBrake window, with its toolbar, Presets drawer, and settings area.

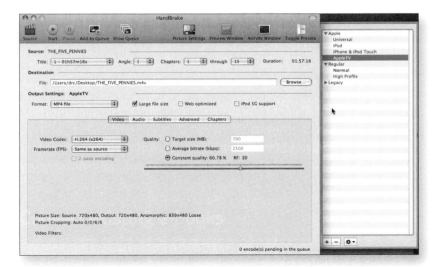

Depending on your preferences settings, HandBrake might display an Open dialog on launch, asking you to specify the source to transcode. If you don't see the Open dialog, click the Source button in the top-left corner of the HandBrake window to make it appear.

 Except for adjusting to minor visual differences, Windows users should be able to follow right along with this project, because HandBrake is pretty much the same on Mac, Linux, and Windows computers.

Getting the software:

1. Download HandBrake at http://handbrake.fr.

2. Download VideoLAN Client at www.videolan.org.

Convert and transfer your media

Making your video available to your iPad is a two-part process. The first part is the actual transcoding operation, which you can deal with by telling HandBrake what your target platform is. The second part is moving the content into iTunes for syncing.

Specifying your transcode format:

1. In HandBrake's Open dialog (refer to "Acquire HandBrake and VLC" earlier in this project), select your input source.

   This source could be a video file on one of your discs, a VIDEO_TS folder on a hard drive, or an inserted DVD. **Figure 3.60** shows an inserted DVD selected.

Figure 3.60
HandBrake displays an Open dialog that lets you navigate to your source content.

2. Click Open.

   The dialog closes, and HandBrake scans the selected source, placing the videos it finds in the Title pop-up menu (**Figure 3.61**).

Figure 3.61 The Title pop-up menu lists the sources found and their durations.

3. Select the desired preset.

   - On a Mac, presets appear in the Presets drawer (refer to Figure 3.59). If the drawer is closed, click Toggle Presets in the toolbar to open it.

   - In Windows, presets appear in a list box on the right side of the HandBrake window.

   Currently, the best preset to choose is AppleTV, but an iPad-specific preset may become available in a new HandBrake version by the time you read this book.

   You can adjust specific settings in the Video, Audio, Subtitles, Advanced, and Chapters panes by clicking the corresponding tabs near the middle of the HandBrake window (refer to Figure 3.59). Unless you know what you mean to accomplish, you probably should leave them alone. We don't want to discourage experimentation in pursuit of knowledge, however, so if you want to try things out, feel free. After all, all you have to lose are some time and (temporarily) some disk space.

Letting it rip:

1. Complete the steps in the preceding section, "Specifying your transcode format," if you haven't already.

   HandBrake displays the path to and name of the destination file.

   If you prefer to save the output to another location, you can type that location in the File text box or click the Browse button to its right and navigate to the new location.

2. Add the transcode to the queue by clicking the Add to Queue toolbar button.

   If you're going to be processing only one file, you can just click the Start button in the toolbar and skip the rest of these steps.

3. Repeat Step 2 for other sources that you want to convert.

   These sources can be other titles already listed in the Title pop-up menu or sources that you add via the steps in "Specifying your transcode format" earlier in this project.

4. Click the Show Queue toolbar button.

   The Queue window opens (**Figure 3.62**).

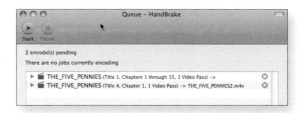

Figure 3.62 The Queue window tells you which sources you've told HandBrake to process.

You can click the disclosure triangle to the left of any job file to see its parameters, as shown in **Figure 3.63**. If you decide that you want to remove a file from the queue, click the gray X button to its right.

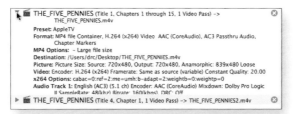

Figure 3.63 Click a queued item's disclosure triangle to see the transcode settings.

5. When you have everything the way you want it, click the Start button at the top of the Queue window.

   An orange circle with two curved arrows chasing each other appears to the left of the job in progress. A check mark in a green circle appears to the left of completed jobs. The X button that appears to the right of pending or in-progress jobs is replaced by a magnifying-glass icon for completed jobs. Click the magnifying glass to see the result in the Finder (Mac) or Windows Explorer.

Moving your content to the iPad:

1. Launch iTunes, if it isn't already running.

2. In the Finder (Mac) or Windows Explorer, locate and select your output file(s).

   See Step 4 of "Letting it rip" earlier in this project for the easiest way to locate your output video.

3. Drag your selected file(s) to the Library section of iTunes' Source list (**Figure 3.64**).

Figure 3.64 Drag your output file(s) to the Library section of iTunes.

By default, your new content syncs to your iPad; if it doesn't, you've changed a default setting somewhere. Go to the Videos tab in iTunes, and make certain that the new content is selected for synchronization.

 If you set up a relationship so that MP4 (or M4V) files "belong" to iTunes, an alternative way to transfer content is to double-click a selection in the Finder or Windows Explorer. iTunes launches and performs the transfer automatically. This relationship seems to be the default on the Mac, but we had to establish the relationship on a Toshiba laptop.

 You can also invoke iTunes' File > Add to Library command and navigate the Add to Library dialog to select the files that you want to add.

Index

A

ABC Player
  basics, 187–189
  favorites, 190–191
  obtaining, 189
Add to Existing Contact button, 28
addresses (email)
  adding contacts from address fields, 27–30
  copying, 32
  editing contacts created from, 30–31
addresses (street)
  adding to contacts, 32
  locating in Maps, 109–111
  locating linked in Maps, 104
ADEPT, 167
air travel
  airports, locating, 142–143
  flight routes, setting, 141–144
  flights, finding, 144–146
Air Video
  acquiring, 197–199
  basics, 197
  user interface, 203
Air Video Server
  acquiring, 197–199
  basics, 197
  Properties dialog for preferences, 202
  setting up, 199–202
alerts (calendars/new invitations), 48–49
Apple. *See also* iWork.com site; Keynote; MobileMe; Pages
  Find My iPhone service, 24–26
  FireWire ports/USB ports, 207
  iBooks, 164–165, 168
artists, selecting for syncing, 155–156
attachments, previewing, 54–55
AVCHD (high-def version of the Advanced Video Codec), 211

B

background color, changing, 94
backing up information, 5
Baen Free Library, 172
Bcc fields, 117
Betamax case, 213
BigOven Lite, 80–86
  emailing recipes from, 86
  leftovers, searching for recipes using, 82–83
  making favorites in, 84–85
  searching in, 80–84
BigOven Pro, 80
bookmarks, syncing, 9–10, 18
borders, adding to maps, 113–114
browsing recipes in Epicurious, 73–76
browsers (Web), finding lost iPads and, 24–25

C

calendars
  alerts, 48–49
  basics of, 46
  default, setting, 46–48
  iCalendar (.ics) format, 35
  subscribing to, 35–36
  syncing, 7–8, 18
  Time Zone Support feature, 49–51

calibre
  acquiring and setting up, 169–172
  basics of, 167
CDs, EyeTV and, 206
check boxes, selecting and unselecting, 155
childproofing, 21–23
cloud services. *See also* Wireless Syncing Project
  from cloud to computers, 13–15
  from cloud to iPad, 16–18
  defined, 11
color
  background color, changing, 94
  font or text color, changing, 94
commercials (TV), 195–196
computers, syncing with. *See also* Movie and TV-Show Syncing Project; Music Syncing Project; Wireless Syncing Project
  connecting to iPads, 4
  moving files from, 54–55
  preventing syncing, 2, 3
  syncing with iPads, 2, 6–7
  transferring files from, 60–62
  transferring Word files to, 65–66
Contact and Calendar Management Project, 44–53
contacts
  adding photos to, 31
  data, adding, 31–32
  editing, 30–32
  getting directions, 51–52
  sharing, 33–34
  sorting, 45–46
  syncing, 6–7, 17
contacts, adding, 27–33
  data, adding to contacts, 31–32
  editing, 30–31
  from email address fields, 27–30
Content Server, 171
Converting Video Project, 213–218
  content, moving to iPads, 218
  HandBrake, acquiring, 214–215
  ripping, 216–217
  transcode format, specifying, 215–216
  VLC, acquiring, 214–215

cooking. *See* iPad Chef Project
copy protection, 198
copying
  email addresses, 32
  images from Web sites, 95
  ingredients into scrapbooks, 92–94
copyright law, 213
Create and Convert E-Books Project, 164–177
  calibre, acquiring and setting up, 169–172
  content, sending to Stanza, 175–177
  e-books, acquiring, 168
  e-books, adding and converting, 172–174
  PDFs, creating, 165–166
customizing
  invitations, 107–108
  recipes in Pages, 94
CutePDF Writer, 166

D

data
  adding to contacts, 31–33
  recovering, 21
date and time zone, setting, 50
dates
  adding movies to iTunes and, 181
  setting travel dates, 144–146
default calendars, setting, 46–48
deleting
  emails, 39–42
  notes, 100
  songs from playlists, 160
dinners, scheduling, 138–141
directions, 49–52
directories. *See* folders (mail)
Documents To Go, 69–71
DRM (digital rights management), 167, 198, 200
Dropbox, 67–69
duplicating slides, 128–129
DVDs
  EyeTV and, 206
  transcoding and, 213, 215

Index

E

e-book readers, 167, 174
editing
  contacts, 30–32
  importing files for, 69–71
  previewed files, 54
  shopping lists in Notes, 103
  slides, 128–129
  text fields, 128
Elgato Systems, 206
emailing
  files to iPad, 54–55
  invitations, 115–117
  notes, 101–102
  presentations, 132
  recipes, 86
  shopping lists, 78–80
emails. *See also* Mail Management Project
  deleting, 39–42
  drafting, 34–35
  moving to folders, 42–44
embedding maps in invitations, 111–114
Epicurious
  displaying favorites, 77–78
  finding recipes with, 72–77
  making favorites in, 77
  navigating, 76
  searching in, 76–77
  viewing and emailing shopping lists, 78–80
Epicurious magazine, iPhone app from, 73
episodes (TV shows)
  available, 192
  syncing and, 183, 184–186
  watching, 190–191
ePub format, 164, 167
eReader, 167, 168
events (calendars)
  displaying/creating, 46–48
  syncing, 8, 13
exporting
  files, 64–66
  flash cards, 131–132
  invitations, 115–116
  presentations to iPad, 132

EyeTV
  Archives option, 209
  capture product, 206
  EyeTV 3 software, 206
  linking, 208
  navigating, 210–212
  Preferences dialog, 208
  sharing preferences, 208

F

fair use, 213
favorites
  ABC's, 190–191
  in BigOven Lite, 84–85
  in Epicurious, 77–78
  file in Dropbox, 67–68
  recipes. *See* scrapbooks for recipes
feedback to ABC, 194
fields (email addresses)
  adding contacts from, 27–30
  editing contacts from, 30–31
File Management Project, 54–72
  Documents To Go, 69–71
  Dropbox, 67–69
  emailing, 54–55
  exporting, 64–66
  importing, 62–63
  opening in Pages, 57–60
  Pages and compatibility, 60
  Phone Disk, 72
  previewing, 55–57
  transferring from computers, 60–62
Find My iPad feature, 23–26
Find My iPhone service, 24
FireWire ports (Macs), 207
Flash Card Project, 119–132
  flash cards, creating. *See* presentations
  flash cards, exporting, 131–132
  Free Translator 50, 120–121
  illustrations, collecting, 123–125
  Keynote, 119–120
  translating words and phrases, 121

flights
  finding, 141–144
  flight info, adding to itineraries, 146–150
  routes, setting, 141–144
folders (mail)
  defined, 27
  moving mail to, 42–44
folders (playlists)
  adding and removing playlists from, 159
  creating in iTunes, 158–159
fonts
  changing in Pages, 94
  substitutions when importing, 58–59
  supported by Pages, 60
formats
  not supported by iPad (video), 196
  for transcoding, 214, 215–216
formatting song size, 154
Free Translator 50, 120–122
full-screen mode, 204–205

G

genie effect, 39
genres (music), selecting for syncing, 155–156
Gerritsen, Tess, 172, 177
Google. See also Maps
  gathering images with, 123–125
  translation services, 120
  wireless syncing and, 12

H

H.264, 211
HandBrake
  acquiring, 214–215
  ripping with, 216–217
  specifying transcode formats, 215–216
hardware
  accessing EyeTV hardware, 210
  needed for Watching Television Project, 206–207

I

iBooks, 164–165, 168
.ics format, 35, 36

illustrations. See also images
  collecting, 123–125
images
  copying from Web sites, 95
  gathering with Google, 123–125
  importing into Pages, 95
  inserting into Pages, 95–97
  saving, 124
IMAP (Internet Message Access Protocol), 26, 37
importing
  files into iPads, 62–63
  Word files into Documents To Go, 69–71
  from Word into Pages, 57–58
in-app purchases, 22
INEPT, 167
Information Syncing Project
  backups, 5
  basics of, 2–3
  calendars, 7–8
  contacts, 6–7
  iTunes for, 3
  notes and bookmarks, 9–10
  settings, applying, 10–11
  settings, mail, 8–9
  settings, viewing, 3–6
ingredients in recipes, copying into scrapbooks, 92
InMethod's Air Video products. See Air Video; Air Video Server
Intel Macintosh, 206
Internet, locating EyeTV software over, 210. See also Web sites for downloading; Web sites for further information
invitations, creating
  addresses, locating with Maps, 109–111
  basics of, 105–108
  list for, creating in Notes, 99–101
  maps, embedding, 111–114
  saving as PDFs, 115–117
iPad
  connecting to computers, 4
  similarity to iPod touch, 152
  space capacity of, 152, 181, 197

Index

iPad Chef Project, 72–97
  iPads, using in the kitchen, 97
  recipe scrapbooks. *See* scrapbooks for recipes
iPhone
  from *Epicurious* magazine, 72
  EyeTV for, 209
iPod app, making playlists with, 161–163
iPod touch, similarity to iPad, 152
italics, formatting, 93
iTunes
  data recovery and, 21
  parental restrictions, 23
  playlists and playlist folders, creating in, 158–159
  restricting use of, 21–22
  for syncing information, 2, 3
  syncing library to iPad, 153
  syncing tabs, 178–179
  transferring Word files via, 60–62, 65–66
  video, moving to, 218
iWork.com site
  sharing presentations via, 132
  transferring files and, 66

J

Jobs, Steve, 164

K

KAYAK
  acquiring, 134
  setting flight routes, 141–144
  setting travel dates and finding flights, 144–146
Keynote
  acquiring, 120
  basics, 119
  creating flash card deck. *See* presentations
Kindle, 167, 168
kitchens, using iPads in, 97

L

languages, changing with Free Translator 50, 121, 130
legality of transcoding, 213
letterbox viewing, 205

library (iTunes)
  syncing to iPad, 153
  videos, adding to, 218
linking software for Watching Television Project, 207–210
links, creating in Notes, 103
lists, using Notes to create, 98–101. *See also* shopping lists
live TV, watching, 211

M

Macs
  creating PDFs on, 165–166
  EyeTV and, 206, 207, 208
  FireWire ports/USB ports, 207
  signing up for MobileMe and, 12, 13
  syncing between MobileMe and computers, 13–14
  syncing calendars on, 7–8
  syncing contacts on, 6–7
  syncing mail on, 9
  syncing notes and bookmarks on, 9–10
  transcoding video on, 216
  transferring files via iTunes on, 62
magnifying glass (Notes), 103
Mail
  attachments previewable by, 55
  editing files in, 54
  launching, 71
  previewing Word files in, 56–57
Mail Management Project, 26–44
  calendars, subscribing to, 35–36
  contacts. *See* contacts, adding
  emails, deleting, 39–42
  emails, drafting, 34–35
  folders, moving mail to, 42–44
  mailbox management, 36–37
  spam, 37–39
mailboxes. *See also* folders (mail)
  managing, 36–37
manual music and video management, 163–164, 178, 180

Maps
  getting directions and, 51–52
  locating addresses with, 109–111
  locating linked addresses, 104
  street addresses and, 32
maps, embedding in invitations, 111–114
masking maps, 113–114
Me button (ABC Player), 193
metadata, for e-books, 167, 173
Microsoft. *See also* Windows; Word, exporting from Pages to; Word files
  Microsoft Exchange for syncing, 12
  PDFs, creating and, 166
MobileMe
  accounts, acquiring, 12–13
  accounts on iPads, creating, 16–18
  features of, 13
  Find My iPad feature, 23–26
  syncing with computers, 13–15
Movie and TV-Show Syncing Project, 178–187
  moving movie rentals, 187
  music videos, 187
  syncing movies, 179–183
  syncing TV-show episodes, 183–186
movie rentals, moving, 187
Music Syncing Project, 152–164
  manual music management, 163–164
  playlists, adding songs to, 159–160
  playlists, creating on iPads, 161–163
  playlists, selecting and syncing, 160–161
  playlists and playlist folders, creating in iTunes, 158–159
  reformatting songs for more space, 154
  smart playlists, 161
  syncing artists and genres, 154–156
  syncing iTunes library to iPad, 153
music videos, syncing, 187

N

New Contact form, 29
Notes
  basics of, 98
  handling RSVPs in, 118
  invitation lists, creating in, 98–101
  shopping lists, creating in, 101–104
notes, syncing, 9–10

O

orientation of iPad
  Air Video user interface and, 203
  Back button in Epicurious and, 75
  Keynote and, 125
  for reading e-books, 177

P

Pages
  compatibility and, 60
  copying ingredients into scrapbooks, 92–94
  creating recipe files in, 90–91
  creating recipe scrapbooks in, 87
  customizing recipes in, 94
  exporting to Word, 64–66
  importing emailed recipes into, 87–90
  importing images into, 95
  importing Word files into, 58–59, 62–63
  inserting images into, 95–97
  invitations, creating in, 105–108
  opening Word files in, 57–60
parental ratings, 22, 23
Party Project, 98–118
  invitations. *See* invitations, creating
  Notes for making lists, 98–101
  responses, 118
  shopping lists, creating, 101–104
passcodes
  passcode locks, 19–21
  restrictions passcode, 22
  setting, 19–20
PDFs
  creating, 165–166
  saving invitations as, 115–117
Phone Disk, transferring files with, 72
phone numbers in contacts, 32
photos. *See also* images
  adding to contacts, 31
  emailing, 35

in invitations, 111–112
Saved Photos, 127, 128
placeholder text (Pages), 93
playback (video), controlling, 195–196, 205
playlists
   adding songs to, 159–160
   Air Video Server and, 200–201
   basics, 156–158
   creating in iTunes, 158–159
   creating on iPads, 161–163
   movies in, 179, 182–183
   playlist folders, creating in iTunes, 158–159
   selecting and syncing, 160–161
   smart playlists, 161
   TV-show episodes and, 185
plus-sign icon, 35
POP (Post Office Protocol) mail accounts, 37
PowerPoint vs. Keynote, 119
presentations, 125–132
   basics of, 125–128
   Italian slides, creating, 130
   slides, duplicating and editing, 128–129
   viewing, 131
previewing
   attachments, 54–55
   files, 54, 55–57
Print dialog to create PDFs, 165
purchases, restricting, 22

Q

queue for processing video, 217

R

readers (e-books), 167, 168, 174
recipes. *See* BigOven Lite; Epicurious; scrapbooks for recipes
recovering
   data, 21
   deleted emails, 39
reformatting song size, 154
responses to invitations, 118
restoring data
   iTunes and, 21
   Restore feature, 26

restrictions, 21–23
ripping video, 216–217
RSVPs, tracking, 118

S

Safari
   collecting illustrations in, 123–125
   disabling and restrictions, 22
   in Flash Card Project, 120
safety of iPads in the kitchen, 97
saving
   email drafts, 34
   files, 71
   images, 124
   invitations as PDFs, 115–117
schedule of TV programs, viewing, 191–192
scrapbooks for recipes, 87–94
   basics of creating, 87
   ingredients, copying into scrapbooks, 92–94
   Pages, importing emailed recipes into, 87–90
   recipe files, creating, 90–91
   recipes, customizing, 94–97
screen shots in Maps, 110–111, 113
searching
   for addresses in Maps, 109–111
   in BigOven Lite, 80–84
   in Epicurious, 76–77
   for flights, 142–144
selecting
   artists/genres for syncing, 155–156
   playlists, 160–161
   selecting and unselecting playlist items, 155
selection buttons, 88, 92–93
settings (for syncing)
   applying, 10–11
   mail, 8–9
   viewing, 3–6
sharing
   contacts, 33–34
   files with Dropbox, 67
shopping lists
   BigOven Pro and BigOven Lite and, 80
   creating in Notes, 101–104
   viewing and emailing, 78–80

shortcuts for addresses and photos (Mail), 35
Show To/Cc Label to check for spam, 38-39
Simple Security Project, 18-26
    childproofing, 21-23
    Find My iPad feature, 23-26
    passcode lock, 19-21
slides
    duplicating and editing, 128-129
    in Keynote, 125
    slide composition screen, 131
smart playlists, 161
software
    for Converting Video Project, 213
    for Create and Convert E-Books Project, 164
    for File Management Project, 54
    for Flash Card Project, 119
    for iPad Chef Project, 72
    for Party Project, 98
    for Streaming Internet Video Project, 187
    for Streaming Your Video Project, 196
    for Vacation Planning Project, 132
    for Watching Television Project, 206
songs
    adding to playlists, 159-160
    references to in playlists, 156-157
    reformatting size, 154
    syncing only favorites, 154-156
Sony Corp v. Universal City Studios, 213
Sort Order screen (contacts), 45
Source list (iTunes), 4
space capacity of iPads, 152, 181, 197
spam, 37-39
Stanza
    advantage of, 168
    sending content to, 175-177
Streaming Internet Video Project, 187-196
    ABC Player basics, 187-189
    ABC Player favorites, 190-191
    ABC Player feedback, 194
    ABC Player, obtaining, 189
    history, viewing, 193
    schedule, viewing, 191-192
    video stream, viewing, 194-196

Streaming Your Video Project, 196-205
    Air Video and Air Video Server, acquiring, 197-199
    Air Video and Air Video Server basics, 197
    Air Video Server, setting up, 199-202
    playing content, 203-205
Summary tab (iTunes), 5
surnames, sorting contacts by, 45
synchronizing, defined, 2
syncing
    defined, 2
    Dropbox and, 67
    between iPad and computers. *See* Information Syncing Project
    iTunes syncing tabs, 178-179
    music. *See* Music Syncing Project
    preventing, 2, 3
    wireless. *See* Wireless Syncing Project

T

templates
    invitation template, 104, 105-106
    for recipe scrapbooks, 89, 91
text
    changing fonts or colors in Pages, 94
    placeholder text (Pages), 93
    text fields, editing, 128
time zone, setting, 50
Time Zone Support feature, 49-51
Toast Titanium for Sonic/Roxio, 206
transcoding video
    Air Video and, 197
    EyeTV and, 210
    format, specifying, 215-216
    legality of, 213
transferring
    files with Phone Disk, 72
    information. *See* syncing
    iWork.com site and, 66
    transferring files via iTunes, 60-62
    Word files to computers, 65-66
translating words and phrases, 120, 121-122
Trash folder, 39
TravelTracker. *See also* KAYAK
    acquiring, 134

add-ons, 138
creating trips, 135-138
flight info, adding to itineraries, 146-150
scheduling dinners, 138-141
TripIt, 135
trips, creating. *See* KAYAK; TravelTracker
Turbo.264 HD, 206, 211, 212
TV shows. *See* Movie and TV-Show Syncing Project; Streaming Internet Video Project; Watching Television Project

U

updating files in Dropbox, 68
URLs in contacts, 32
USB ports (Macs), 207

V

Vacation Planning Project, 132-150
.vcf (vCard) files, attaching to emails, 33-34
video. *See also* Converting Video Project; Streaming Internet Video Project
  manual management of, 163-164, 178, 180
  music videos, 187
VideoLAN Client (VLC), 214-215
viewing controls for TV-show playback, 195-196
viewing history (TV shows), reviewing, 193
viewing video stream, 194-196
VLC (VideoLAN Client), 214-215

W

Watching Television Project, 206-212
  EyeTV, navigating, 210-212
  linking, 207-210
  software and hardware for, 206-207
Web browsers, finding lost iPads and, 24-25
Web sites for downloading
  Air Video and Air Video Server, 197
  Baen Free Library, 172
  calibre, 169
  CutePDF Writer, 166
  Dropbox software, 67
  Google images, 123
  HandBrake, 215
  KAYAK, 134

Phone Disk, 72
TravelTracker, 134
VLC, 215
Web sites for further information
  Dropbox accounts, 67
  Elgato Systems, 207
  fonts supported by Pages, 60
  Free Translator 50, 120
  Google Sync page, 12
  iCalShare and iCal World, 35
  Keynote, 120
  POP vs. IMAP, 37
  restoring data, 21, 26
welcome wizard (calibre), 170
Windows
  converting video and, 215, 216
  PDFs, creating and, 166
  signing up for MobileMe and, 12-13
  syncing between MobileMe and computers, 13, 14-15
  syncing calendars on, 7-8
  syncing contacts on, 6
  syncing mail on, 9
  syncing notes and bookmarks on, 9-10
  wired syncing, turning off, 15-16
Wireless Syncing Project, 11-18
  cloud services, 11
  cloud to iPad, 16-18
  computers to cloud, 13-15
  MobileMe and, 11
  MobileMe account, acquiring, 12-13
  syncing MobileMe with computers, 13
  wired syncing, turning off, 15-16
wizards
  Getting Started Wizard (Dropbox), 69
  welcome wizard (calibre), 170
Word, exporting from Pages to, 64-66
Word files
  importing into Pages, 62-63
  opening in Pages, 57-60
  previewing in Mail, 56-57
  transferring to computers, 65-66
  transferring via iTunes, 60-62

WATCH READ CREATE

Meet Creative Edge.

A new resource of unlimited books, videos and tutorials for creatives from the world's leading experts.

Creative Edge is your one stop for inspiration, answers to technical questions and ways to stay at the top of your game so you can focus on what you do best—being creative.

All for only $24.99 per month for access—any day any time you need it.

peachpit.com/creativeedge